Cognitive Choice Modeling

Cognitive Choice Modeling

Zheng Joyce Wang and Jerome R. Busemeyer

The MIT Press
Cambridge, Massachusetts
London, England

This book was set in Stone Serif and Stone Sans by Westchester Publishing Services. Printed and bound in the United States of America.

Library of Congress Cataloging-in-Publication Data

Names: Wang, Zheng Joyce, author. | Busemeyer, Jerome R., author.
Title: Cognitive choice modeling / Zheng Joyce Wang and Jerome R. Busemeyer.
Description: Cambridge, Massachusetts : The MIT Press, [2021] | Includes bibliographical references and index.
Identifiers: LCCN 2020019868 | ISBN 9780262044967 (hardcover)
Subjects: LCSH: Choice (Psychology)—Mathematical models. | Decision making—Mathematical models.
Classification: LCC BF611 .W36 2021 | DDC 153.8/3011—dc23
LC record available at https://lccn.loc.gov/2020019868

10 9 8 7 6 5 4 3 2 1

This book was a labor of love for Angela

Contents

Preface

Cognitive decision processes provide the interface between the environment and brain, enabling choice behavior. The basic cognitive mechanisms underlying decision processes are fundamental to all fields of human activity. Yet cognitive processes and choice processes are often studied separately, whether by decision theorists, consumer researchers, or other behavioral and social scientists. In this book, we introduce a new cognitive modeling approach to the study of human choice behavior. Integrating recent research from both cognitive science and choice behavior, the book provides the groundwork for the emerging interdisciplinary field of cognitive choice modeling.

The focus of this book is on individual choice behavior. It begins (chapter 1) with an introductory review of the basic mathematical background needed to understand for the remaining chapters. The content requires an undergraduate background in matrix algebra and calculus, and a beginning graduate-level background in statistics.

The next two chapters (chapters 2–3) provide a survey of traditional decision science covering utility theory and random utility models. Utility theory (chapter 2) is formulated in a deterministic framework, but human behavior is less predictable. Random utility models (chapter 3) provide a probabilistic reformulation of utility models, and these models provide the backbone for traditional choice modeling books. We review both basic theory and empirical research into these two classes of models.

What makes this book stand out are the chapters that cover theory and empirical research originating from cognitive science. This part of the book begins with chapter 5 on one of the oldest theories, called signal detection theory, which remains influential today, as well as its newer, multivariate extensions. Signal detection theory is used to decide an action based on a sample of evidence about an unknown state of nature. The theory is designed to account for both choice probability and confidence in choice behavior. However, the theory is a static theory, and thus it fails to describe the dynamic process that produces a choice.

The next three chapters (chapters 6, 7, and 8) survey the theory and research on sequential sampling models, which describe and predict not only choice and confidence, but also the time it takes to make a choice. These chapters cover applications of sequential sampling models to both evidence- and value-based decisions. When applied to evidence-based decisions, sequential models provide dynamic extensions of signal detection theory. When applied to value-based decisions, sequential sampling models provide dynamic extensions of random utility theories.

While sequential sampling models are useful for describing the dynamics of a single choice, they don't describe the dynamics across a sequence of choices produced by learning from experience. For this reason, chapter 9 describes the theory behind and research on reinforcement learning models, reviews research comparing description- versus experience-based decisions, and discusses new theoretical work comparing model-free and model-based learning theories.

After presenting the behavioral and cognitive foundations of choice, in chapter 10, we provide an overview of the progress being made toward understanding the relations between cognitive choice models and the neural mechanisms underlying choice behavior. We also examine the body of evidence supporting the neural substrates that underly utility models, sequential sampling models, and reinforcement learning models.

Finally, in chapter 11, we examine an entirely new and different perspective for modeling choice behavior based on quantum probability theory. In many ways, quantum probability provides a generalization of classical probability. It may be needed to account for findings that seem puzzling from a classical probability point of view.

Another important chapter, that is chapter 4, concerns the statistical estimation of model parameters and methods for comparing competing models. It comes right after we discuss random utility models, so we can present a detailed example of applying random utility models to real experimental choice data. In this chapter, we review traditional maximum likelihood methods, as well as more recent hierarchical Bayesian methods.

The book ends with chapter 12, which provides some speculation about the directions of future research on cognitive choice modeling.

The book was organized with a theme that builds from simple deterministic utility theories to more sophisticated static random utility theories, and then moves to examining more complex stochastic sequential sampling and reinforcement learning processes. However, the chapters were written in a way that does not necessarily require reading earlier chapters, and each chapter can be read somewhat independent of the others.

Acknowledgments

This book has benefited from valuable comments from a number of colleagues working in the field of cognitive choice modeling, including Woo-Young Ahn, Sudeep Bhatia, Ryan Jessup, Ian Krajbich, Emmanual Pothos, Neil Stewart, Jennifer Trueblood, and Brandon Turner. We greatly appreciate their help to improve the accuracy and readability of the book. Part of the work was supported by the National Science Foundation (NSF SES-1560554, SES-1560501).

1 Introduction

Choice behavior is a topic that should be of interest to all researchers who study human behavior. People make decisions every moment of their lives, both large and small, whether it is choosing a word that communicates the right ideas to another person, choosing the color of a shirt to wear to work, choosing a new car, choosing a medical insurance plan for an elderly family member, or choosing a course of military action. This broad appeal makes the topic very interdisciplinary, ranging across psychology, consumer economics, communication, sociology, political science, and other fields. Likewise, many excellent books on choice behavior have been written by economists, consumer researchers, and transportation scientists, among other experts (e.g., Ben Akiva & Lerman, 2017; Hess & Daly, 2014; Train, 2003), but this book is different. We will present here what we call a *cognitive modeling* approach to human choice behavior. Although we cover many of the traditional topics concerning choice behavior, we also introduce new ideas arising primarily from recent research by cognitive scientists.

Our goal is to provide a comprehensive survey that covers both the basic empirical findings concerning choice behavior and the choice models that have been proposed to account for these findings. A rigorous presentation of choice models requires some mathematical skill in probability theory and statistics, as well as computer programming skills for parameter estimation and model comparison. We do plan to present some background on mathematics and computer programming and to provide a tutorial on some of the recent developments in model comparison methods. However, these topics are secondary to our main goal, which is presenting the basic behavioral findings and theoretical ideas about the concept of choice.

1.1 Overview of Topics

Any book on choice behavior must begin with an introduction to utility theory, including the multiattribute weighted additive utility model and expected utility theory. These utility models often provide the core of many choice models. Our survey begins in chapter 2 with theory that does not come from cognitive science, but rather from decision science.

Next, in chapter 3, we cover random utility models. Choices tend to vary across people, and even across occasions for the same person. Therefore, a choice model must be capable of predicting the probability that a person makes a particular choice. Utility models are deterministic, and in this form, they cannot be applied directly to choice data without making additional assumptions. Random utility models are designed to extend utility models to account for the probabilistic nature of choice behavior.

During our discussion of utility models, we survey some of the behavioral findings that challenge these traditional models. These findings are the culmination of over fifty years of empirical research on choice behavior by psychologists, behavioral economists, and consumer researchers. The entire pattern of findings provides a puzzling array of results that appear paradoxical from the point of view of traditional choice models. These findings represent the challenge that any cognitive model must strive to address.

Chapter 4 is primarily about methodology. Here, we present statistical methods for estimating and comparing choice models, including maximum likelihood estimation and Bayesian estimation and the associated methods for comparing models. We also discuss the advantages and disadvantages of fitting models to aggregate versus individual data, as well as the compromise of using hierarchical mixture models. We provide a detailed example of model estimation and testing using both maximum likelihood methods and hierarchical Bayesian methods, using results from a real empirical study, and using a fairly simple random utility models to illustrate the methods.

After presenting these basic theoretical and empirical ideas related to utility theory, we start to introduce theories in chapters 5 and 6 that have been developed more centrally by cognitive psychologists, including signal detection theories and sequential sampling theories of decision-making. Signal detection theories are designed to model choices as well as confidence ratings, and these theories provide measures of the decision-maker's sensitivity and bias. However, signal detection theories are static theories that are unable to describe the time course of making a decision. Sequential sampling theories go an important step further, providing models for determining not only choice and confidence, but also decision time. These theories describe the time course of decision-making, as well as providing measures of discriminability, bias, and the amount of evidence required to reach a decision.

Initially, sequential sampling theories were applied only to evidence-based decisions (e.g., choices based on accumulating evidence). Later theories were developed and applied to value-based decisions (e.g., choices based on accumulating preferences). Chapters 7 and 8 review more recent dynamic and connectionist network models of preferential choice behavior proposed by cognitive scientists. Chapter 7 reviews a theory from one of the authors, called *decision field theory,* which was one of the first sequential sampling models applied systematically to preferential and risky choice phenomena. Chapter 8 reviews many other theories that have been proposed more recently and compares and contrasts all the competing theories.

None of these theories mentioned above addresses issues related to learning from previous choices. However, choices are often based on experience with consequences in the past, and in chapter 9, we discuss theoretical work on the topic of integrating learning and choice. This chapter introduces theories of reinforcement learning, describes a detailed example of a popular experience-based choice task, and then details the differences between model-free and model-based learning theories.

All the preceding chapters address only behavioral measures. However, a great deal of research has been conducted using electrophysiological measures of the brain and functional magnetic resonance imaging (fMRI) of the brain during decision-making. As a result, great progress has been made toward understanding the neural mechanisms that underly the decision-making process described by the cognitive models. In chapter 10, we present the rapid but impressive progress that has been made toward understanding the relations between cognitive choice models and the neural bases of choice behavior.

Chapter 11 introduces quantum choice models, an entirely new way of thinking about the probabilistic nature of choice. All the previous models of choice essentially assume that choice is determined by a state that existed immediately before the choice is made. But quantum choice models propose that the decision-maker remains in an indeterminate (known as *superposed*) state immediately before a decision, and the choice is selected by an inherently probabilistic process. This chapter provides an introduction to quantum cognition theory and describes an example of quantum probability being applied to risky decision-making.

Finally, in chapter 12, we review the advances that have been made and critically evaluate the various theories that we have discussed in the book. In addition, we make predictions about the directions for future research.

1.2 Justification for Using Cognitive Choice Models

So why bother with using cognitive choice models when there are already good econometric discrete choice models that one could use? For cognitive scientists, the answer is easy—we wish to understand the fundamentals, both the psychological and brain mechanisms, that underly decision-making processes. By doing this, the same cognitive mechanisms can be used to simultaneously predict several different behaviors (e.g., choice, confidence, decision time), as well as neural measures (e.g., electrophysiological activity, fMRI bold signals), using a single common set of parameters. The simultaneous prediction of several measures (e.g., choice and decision time) provides stronger methods for testing and comparing competing models. Furthermore, the model parameters provide meaningful measures of the underlying psychological and neural processes.

This approach may not be satisfactory for an applied consumer researcher. A study that is more interested in application (e.g., making profits) may be more difficult to perform successfully because the major purpose for an application may simply be making predictions rather than reaching a deeper scientific understanding. Just build a model with a sufficient number of parameters that is needed to fit the data. For example, if a particular context has an unexpected effect on choice, simply add a new parameter whenever that context is present. But does blind addition of parameters really work? Adding a new parameter for every unexpected new effect may provide good *postdiction,* but not really *prediction*. In other words, adding another parameter may improve the model fit after seeing the data, but it could still fail to predict behavior in new contexts. We argue that the most important added value of a cognitive model is to provide a priori predictions in new situations. A good cognitive model can be used to forecast how a change in context will affect choice. For example, a cognitive model could predict the effect that adding a new product would have on an existing market, and by doing this, help design that product to maximize the desired effect. We will show in later chapters that in fact, cognitive models perform better than traditional econometric models for predicting context effects on choice.

In sum, choice models from cognitive science may appear to be more complicated than traditional choice models by consumer economists. Therefore, throughout this book, our task is to demonstrate the need and the added value of using cognitive models rather than the more traditional consumer choice models. We will attempt to show that cognitive models provide the ability to account for a wider range of behavioral (and neural) phenomena in a more consistent and parsimonious manner than traditional choice models.

1.3 Focus of This Book

This book is about choice behavior, but what exactly do we mean by that? More precisely, what do we include and what do we leave out? After all, choices can be made by individuals (e.g., the president of the US) or groups (e.g., a Senate Committee on Finance); choices can be made alone (e.g., a teenager deciding whether to commit suicide) or in a strategic interaction with one or more parties (e.g., making a business deal between US and Chinese companies); and choices can be made in a single step (e.g., buying a new car) or dynamically across time in a sequence of interdependent steps (e.g., planning a military campaign). Although group decision-making is an interesting topic and game theory is a very important area of research, we will concentrate here on *individual choice behaviors*. Also, although most real-life choices are only one step in a dynamic sequence of decisions, we will primarily consider choices made in a single, stand-alone step. It is not that we think individual choice in a single step is most important, but rather that we need to limit our focus, and the theories that arise in this simple setting are likely to serve as a foundation for more complex scenarios as well.

1.4 Basic Statistical Concepts

Choice models and the statistical analysis of these models require knowledge of several basic mathematical statistical concepts. We review the necessary ideas next; see Hogg and Craig (1970) for more details on mathematical statistics.

Probability distributions of random variables Choice models are almost always probabilistic in nature. This is because the data obtained from choice studies usually concern the relative frequencies of events. These relative frequencies are assumed to be sample estimates generated by some probabilistic choice mechanism. In some applications, the frequencies are obtained by aggregating responses across individuals from a population of interest. In other cases, the relative frequencies are obtained by pooling across choice trials for a single individual.

The abstract definition of a probability distribution is based on the following three simple axioms formulated by Kolmogorov (1932). Suppose that S is a set that contains all the possible events that can occur, and that the events A and B are subsets of S. The event (A or B) is represented as the union of the two subsets, and the event (A and B) is represented by the intersection of the two subsets.

The first axiom states that the probability assigned to an event is a positive number. The second axiom states that the probability of the event S (which contains anything that can happen) is 1, so the probabilities range between zero and 1. The third axiom states that if the pair A, B are mutually exclusive, then the probability of A or B is the probability of A plus the probability of B. (This concept is called *finite additivity*, and although a more demanding sigma-additive assumption is often needed, this is sufficient for this discussion.) The third axiom is quite powerful. In particular, it tells us how to assign probabilities to the union of events when they are not mutually exclusive: $p(A \text{ or } B) = p(A) + p(B) - p(A \text{ and } B)$. The third axiom also implies that the probability of the null or impossible event (e.g., A and ~A) is zero, because $p(S \text{ or } {\sim}S) = p\ (S) + p({\sim}S) = 1$.

The last important concept is that of *conditional probability*. Define $p(A$ and $B)$ as the joint probability that A and B both occur. Then the probability of event B given that event A occurs is defined as $p(B \mid A) = p(A \text{ and } B)/p(A)$.

One might wonder how this abstract definition is interpreted in choice modeling. There are at least three ways to interpret probabilities: relative frequency, propensity, and subjective probability. For example, in the case of choice studies that estimate a relative frequency of an event by aggregating across a sample of individuals from a finite population, we might interpret the probability of the event as the population proportion that would select that event. In another example, in the case of choice studies obtaining a sequence of responses from a single individual, we might interpret the probability of an event as the propensity or tendency for that individual to select that event. As a final example, when testing a hypothesis about a choice mechanism, we might have a prior belief in the hypothesis that is updated (using the conditional probability formula) into a posterior belief by the relative frequency data that we observe. Thus, all three of these interpretations may occur in choice modeling, depending on the application.

When a researcher takes a measurement (by assigning a number to an event), we say that a random variable is produced. A *random variable* is a function that assigns real numbers to events. For example, suppose that a person rates the confidence that an answer is correct on an 11-point scale. In this case, we can define a random variable X to represent the confidence rating procedure that is used to assign possible numbers (ranging from 1 to 11) to the rating categories that could be selected by a person. We assign a probability to each value of the random variable that can be observed. For example, $p(X=5)$ stands for the probability of the event that the person picks the value 5. Now suppose that this random variable has N values $\{x_1, x_2, \ldots, x_i, \ldots, x_N\}$ (previously $N=11$ scale values). The function, $p(X=x_i)$, which stands for the probability that the random variable X is assigned the number x_i, is called the *probability mass function* for the random variable X. The cumulative probability is then defined as $p(X \leq x_i)$, which is the sum of the probabilities of each value of X that is less than or equal to x_i.

Often, we measure more than one random variable. For example, we could observe the age of a person and the number of hours per week that the person exercises. Suppose that X is the random variable for age (in years), and let Y be the random variable for hours of exercise. Then we can define a joint probability $p(X=x_i \cap Y=y_j)$, which equals the probability that x_i and y_j occur. Finally, we can define the conditional probability of finding $Y=y_j$, given that we have observed $X=x_i$ as the ratio

$$p(Y=y_j \mid X=x_i) = \frac{p(X=x_i \cap Y=y_j)}{p(X=x_j)}.$$

Often, we work with random variables that have a continuous rather than a discrete and finite distribution. For example, the decision time T required to make a choice is approximately continuous. Suppose that X is a continuous random variable. In this case, the probability assigned to each value on the continuum has zero measure (there is an uncountably infinite number of them in any interval). So instead, we start by defining the cumulative distribution function $F(x)=p(X \leq x)$. Assuming that the cumulative distribution is sufficiently smooth, then we can define the probability density at each value of x as the derivative of the cumulative probability function, $f(x)=dF(x)/dx$. Then $F(x)=\int_{-\infty}^{x} f(x)dx$, the integral of the density.

Moments of the distribution When working with random variables, we are often interested in their *moments,* such as their means or variances or correlations. These are all defined as expected values of random variables. The expected value of a random variable is defined as

$\mathrm{E}(X)=\Sigma\, p(X=x_i)\cdot x_i$ for the discrete case and

$\mathrm{E}(X)=\int f(x)\cdot x dx$ for the continuous case.

The mean of a random variable X, denoted as μ_X, is defined as the expectation of the random variable $\mu_X=\mathrm{E}(X)$. The variance of a random variable X, denoted as σ_X^2, is defined as the expectation of the squared deviation around the mean as follows:

$$\sigma_X^2=\mathrm{Var}(X)=\mathrm{E}(X-\mu_X)^2.$$

For example, in the discrete case, this equals

$$\mathrm{Var}(X)=\Sigma\, p(X=x_i)\cdot(x_i-\mu_X)^2.$$

The standard deviation is the square root of the variance, $\sigma_X=\sqrt{\sigma_X{}^2}$. The covariance between two random variables (X, Y), denoted as σ_{XY}, is defined by the expectation of the product of deviations as follows:

$$\sigma_{XY}=\mathrm{Cov}\,(X, Y)=\mathrm{E}[(X-\mu_X)\cdot(Y-\mu_Y)].$$

For example, in the discrete case, this equals

$$\sigma_{XY}=\Sigma_i\Sigma_j\, p(X=x_i \cap Y=y_j)[(X-\mu_X)\,(Y-\mu_Y)].$$

The correlation is defined as

$$\rho_{XY}=\sigma_{XY}\,/(\sigma_X\cdot\sigma_Y).$$

If we have a set of p random variables, $\{X_1, X_2, \ldots, X_p\}$, we can form a $p\times 1$ random vector $\boldsymbol{X}=[X_1, X_2, \ldots, X_p]$. The mean of this random vector is a $p\times 1$ vector:

$$\mathrm{E}(\boldsymbol{X})=[\mathrm{E}(X_1), \mathrm{E}(X_2), \ldots, \mathrm{E}(X_p)]=[\mu_1, \mu_2, \ldots, \mu_p]=\boldsymbol{\mu}.$$

The random vector also has a variance—covariance matrix, which is a $p \times p$ matrix:

$$\Sigma = \text{Var}(\mathbf{X}),$$

with σ_{ij} in row i and column j of Σ, where the ith diagonal entry $\sigma_{ii} = \text{Var}(X_i)$ equals the variance of the random variable X_i, and $\sigma_{ij} = \text{Cov}(X_i, X_j)$ for row $i \neq$ column j is the covariance between random variables X_i and X_j. We describe how to work with vectors and matrices in the next section.

1.5 Basic Matrix Algebra

Our presentation of choice models in this book also requires some basic knowledge of matrix algebra (Strang, 1988). This topic is important for any kind of multivariate statistical analysis.

An $n \times m$ matrix is a table of numbers that has n rows and m columns, which is denoted by a capital letter such as **X**. The indices, n and m, define the dimensions of the matrix. Here is an example of a 2 (row) $\times$ 3 (column) matrix:

$$\mathbf{X} = \begin{bmatrix} 1 & 0 & 3 \\ 2 & 1 & 4 \end{bmatrix}.$$

The transpose of a matrix, symbolized by $\mathbf{X}^{\mathrm{T}}$, changes the columns into rows:

$$\mathbf{X}^T = \begin{bmatrix} 1 & 2 \\ 0 & 1 \\ 3 & 4 \end{bmatrix}.$$

A column vector can be viewed as a matrix with only one column:

$$\mathbf{B} = \begin{bmatrix} 1 \\ 0 \\ 2 \end{bmatrix}.$$

A row vector is a matrix with only one row:

$$\mathbf{C} = [\, 3 \quad 1 \quad -1 \,].$$

A row vector can be viewed as the transpose of a column vector:

$$\mathbf{B}^T = [\, 1 \quad 0 \quad 2 \,].$$

It is often more convenient to use row vectors inside the text of a document.

A square $n \times n$ matrix is a matrix with the same number of rows and columns, which we simply say has dimension n. A matrix with zeros in the off diagonals and nonzero entries on the diagonals is called a *diagonal matrix*. A special matrix, called the *identity matrix* and denoted as **I**, is a diagonal matrix with 1s in the diagonal entries and zeros in all the off-diagonal entries. For example, note the case of the following three-dimensional matrix:

$$\mathbf{I} = \begin{bmatrix} 1 & 0 & 0 \\ 0 & 1 & 0 \\ 0 & 0 & 1 \end{bmatrix}.$$

Two matrices can be summed if they have the same dimensions, and the matrix sum is obtained by summing the corresponding elements. For example, define another matrix as

$$\mathbf{D} = \begin{bmatrix} 1 & 2 & 1 \\ 2 & 1 & 2 \end{bmatrix}.$$

Then, using the previous definition of **X**,

$$\mathbf{X} + \mathbf{D} = \begin{bmatrix} 1 & 0 & 3 \\ 2 & 1 & 4 \end{bmatrix} + \begin{bmatrix} 1 & 2 & 1 \\ 2 & 1 & 2 \end{bmatrix} + \begin{bmatrix} 2 & 2 & 4 \\ 4 & 2 & 6 \end{bmatrix}.$$

Scalar multiplication is performed by multiplying a single number by each element of a matrix, such as

$$2 \cdot \mathbf{X} = \begin{bmatrix} 2 \cdot 1 & 2 \cdot 0 & 2 \cdot 3 \\ 2 \cdot 2 & 2 \cdot 1 & 2 \cdot 4 \end{bmatrix} = \begin{bmatrix} 2 & 0 & 6 \\ 4 & 2 & 8 \end{bmatrix}.$$

A new concept that arises with matrix algebra is that of an inner product. An *inner product* is computed from a row and column vector of the same dimension. We will denote the inner product by the notation $\mathbf{C} \cdot \mathrm{B}$. The inner product is defined as the sum of the cross-products of the coordinates for the two vectors. For example, using the definitions of **C** and **B** given previously,

$$\mathbf{C} \cdot \mathbf{B} = [\,3 \quad 1 \quad -1\,] \cdot \begin{bmatrix} 1 \\ 0 \\ 2 \end{bmatrix} = (3)(1) + (1)(0) + (-1)(2) = 1.$$

The inner product is used as a measure of similarity. In fact, the correlation is based on an inner product. If the inner product between two vectors is zero, then they are orthogonal. The inner product can also be used to define the squared length of a vector, such as $\|\,\mathbf{B}\,\|^2 = \mathbf{B}^{\mathrm{T}} \cdot \mathbf{B} = 3^2 + 1^2 + (-1)^2 = 11$. If a vector has a length equal to 1, then it is normalized.

The matrix product is another new concept in matrix algebra. A *matrix product* is the product of two matrices. It is defined by a matrix of inner products. If we multiply an $n \times m$ matrix **X** by an $m \times p$ matrix **Y**, then the answer is an $n \times p$ matrix **Z**. The element in the *i*th row and *j*th column of **Z** is the inner product of the *i*th row of **X** and the *j*th column of **Y**. For example, using the previously defined matrices,

$$\mathbf{X} \cdot \mathbf{B} = \begin{bmatrix} 1 & 0 & 3 \\ 2 & 1 & 4 \end{bmatrix} \cdot \begin{bmatrix} 1 \\ 0 \\ 2 \end{bmatrix} = \begin{bmatrix} (1)(1)+(0)(0)+(3)(2) \\ (2)(1)+(1)(0)+(4)(2) \end{bmatrix} = \begin{bmatrix} 7 \\ 10 \end{bmatrix}.$$

Note that we cannot produce a matrix product $\mathbf{X} \cdot \mathbf{D}$ because **X** has three columns and **D** has only two rows. However, following up on the previous example, we can form the matrix product

$$\mathbf{X}^T \cdot \mathbf{D} = \begin{bmatrix} 1 & 2 \\ 0 & 1 \\ 3 & 4 \end{bmatrix} \cdot \begin{bmatrix} 1 & 2 & 1 \\ 2 & 1 & 2 \end{bmatrix} = \begin{bmatrix} 5 & 4 & 5 \\ 2 & 1 & 2 \\ 11 & 10 & 11 \end{bmatrix}.$$

In general, matrix multiplication is not commutative (e.g., $\mathbf{X}^T \cdot \mathbf{D}$ does not equal $\mathbf{D} \cdot \mathbf{X}^T$, even though both exist), and so the order in which the matrices are multiplied matters. It is important to keep the order. The identity matrix has the following special property: $\mathbf{X} \cdot \mathbf{I} = \mathbf{X} \cdot \mathbf{I} = \mathbf{X}$; in other words, it leaves **X** unchanged.

The inverse of a matrix is used to perform matrix division. The inverse of the square matrix **Y**, if it exists, is another square matrix, which we denote as $\mathbf{Y}^{-1}$. The matrix inverse is defined by the property that $\mathbf{Y}\mathbf{Y}^{-1} = \mathbf{I} = \mathbf{Y}^{-1}\mathbf{Y}$. For example, if we define **Y** as

$$\mathbf{Y} = \begin{bmatrix} 2 & 1 \\ 5 & 5 \end{bmatrix}, \text{then } \mathbf{Y}^{-1} = \begin{bmatrix} 1 & -.2 \\ -1 & .4 \end{bmatrix}, \text{because} \begin{bmatrix} 2 & 1 \\ 5 & 5 \end{bmatrix} \cdot \begin{bmatrix} 1 & -.2 \\ -1 & .4 \end{bmatrix} = \begin{bmatrix} 1 & 0 \\ 0 & 1 \end{bmatrix}.$$

The inverse can be used to perform matrix something analogous to division as follows. Suppose that **Y** is defined as done previously, and

$$\mathbf{Z} = \begin{bmatrix} 5 \\ 3 \end{bmatrix},$$

and in addition, we wish to solve for **W** in the equation $\mathbf{Z} = \mathbf{Y} \cdot \mathbf{W}$. Then the solution is

$$\mathbf{Y}^{-1}\mathbf{Z} = \mathbf{Y}^{-1}\mathbf{Y} \cdot \mathbf{W} = \mathbf{I} \cdot \mathbf{W} = \mathbf{W},$$

$$\mathbf{W} = \mathbf{Y}^{-1} \cdot \mathbf{Z} = \begin{bmatrix} 1 & -.2 \\ -1 & .4 \end{bmatrix} \cdot \begin{bmatrix} 5 \\ 3 \end{bmatrix} = \begin{bmatrix} 4.4 \\ -3.8 \end{bmatrix}.$$

The columns of a matrix are said to be linearly independent if it is not possible to derive one column from a linear combination of the remaining columns. In that case, the matrix is said to be *full rank*, and an inverse of the matrix will exist. If one of the columns is linearly dependent on others, then the matrix is less than full rank and the inverse does not exist. (There is something called a *generalized inverse* for the latter case.)

Two final topics that we will use later in this book are the concepts of eigenvectors and eigenvalues. If $\mathbf{X}=\mathbf{X}^T$, then the matrix $\mathbf{X}$ is symmetric: the rows are the same as the columns, such that x_{ij} in row i and column j is the same as x_{ji} in row j and column i. For example, a variance-covariance matrix of a random vector is symmetric. In the following example, suppose that $\mathbf{X}$ is a symmetric matrix.

Consider a vector $\mathbf{V}_j$ that satisfies the linear equation $\mathbf{X}\cdot\mathbf{V}_j=\lambda\cdot\mathbf{V}_j$. Then $\mathbf{V}_j$ is an eigenvector of $\mathbf{X}$, and λ_j is the corresponding eigenvalue. In fact, if $\mathbf{X}$ is an n-dimensional symmetric matrix, then there are n eigenvalues (some may be repeated) and n orthonormal eigenvectors (they may not all be unique unless no eigenvalues are repeated). Furthermore, we can decompose the matrix $\mathbf{X}$ as follows: define the n-dimensional matrix $\mathbf{V}$ as a matrix with n columns and column j is one of the eigenvectors $\mathbf{V}_j$, and define Λ as a diagonal matrix with λ_j in the ith diagonal entry. Then the matrix $\mathbf{X}$ can be decomposed into the product $\mathbf{X}=\mathbf{V}\cdot\Lambda\cdot\mathbf{V}^T$. This is called the *spectral decomposition* of the matrix.

To practice what we have reviewed, let us consider an important application that combines what we have learned about statistical theory with matrix algebra. Suppose that $\mathbf{X}$ is a random $p\times 1$ vector with a $p\times 1$ mean vector $\mathrm{E}(\mathbf{X})=\mu$. Using matrix notation, the variance-covariance matrix is defined as the expectation

$$Var(\mathbf{X}) = E((\mathbf{X}-\mu)\cdot(\mathbf{X}-\mu)^T).$$

What this tells us to do is to first form a $p\times p$ outer product matrix $(\mathbf{X}-\mu)\cdot(\mathbf{X}-\mu)^T$ formed by multiplying a $p\times 1$ column vector by a $1\times p$ row vector. Then we take the expected value of each element of this outer product matrix. The element in row i and column j is simply the product $(X_i-\mu_i)\cdot(X_j-\mu_j)$, and the expected value of this product simply is $\mathrm{E}((X_i-\mu_i)\cdot(X_j-\mu_j))=\mathrm{Cov}(X_i, X_j)=\sigma_{ij}$.

Now suppose that we form a new $q\times 1$ random vector $\mathbf{Y}$ by a linear transformation of the original random vector $\mathbf{X}$ using a $q\times p$ matrix $\mathbf{M}$ as follows: $\mathbf{Y} = \mathbf{M} \cdot \mathbf{X}$. Now we wish to find the mean and the variance covariance matrix of the new vector $\mathbf{Y}$. The mean of $\mathbf{Y}$ is $\mathrm{E}(\mathbf{M} \cdot \mathbf{X}) = \mathbf{M} \cdot \mathrm{E}(\mathbf{X}) = \mathbf{M} \cdot \mu$. The variance—covariance matrix of the new vector $\mathbf{Y}$ equals

$$\begin{aligned} Var(\mathbf{Y}) &= E((\mathbf{M} \cdot \mathrm{X} - \mathbf{M} \cdot \mu) \cdot (\mathbf{M} \cdot \mathrm{X} - \mathbf{M} \cdot \mu)^T) \\ &= E(\mathbf{M} \cdot (\mathrm{X} - \mathbf{M} \cdot \mu) \cdot (\mathbf{M}(\mathrm{X} - \mathbf{M} \cdot \mu))^T) \\ &= E(\mathbf{M} \cdot (\mathrm{X} - \mathbf{M} \cdot \mu) \cdot (\mathrm{X} - \mathbf{M} \cdot \mu)^T \cdot \mathbf{M}^T) \\ &= \mathbf{M} \cdot \mathrm{E}((\mathrm{X} - \mathbf{M} \cdot \mu) \cdot (\mathrm{X} - \mathbf{M} \cdot \mu)^T) \cdot \mathbf{M}^T \\ &= \mathbf{M} \cdot Var(\mathrm{X}) \cdot \mathbf{M}^T \end{aligned}$$

The last line here is an important formula for the variance of a linear transformation, which is used frequently in multivariate analyses. This completes our elementary review of basic mathematical concepts. In this book, we will be using these concepts to describe various choice models or to perform statistical analyses of these models.

1.6 Computer Programming

Cognitive choice models are substantive models that are specially designed for various applications. Therefore, there is no generic, general-purpose computer package that can be used for these models. Instead, researchers have to either download a specific application from a developer's website if it is available, or program the model themselves. Usually, if a new model is being developed, the latter must be done. In that case, programs such as Mathematica, MATLAB, R, and Python are used to compute the model predictions. Most of the previously developed models that we will describe are available for download on websites mentioned in the relevant chapters. All the programs that we wrote for this book are available on our website https://github.com/jbusemey/ChoiceModeling.git. Also, we present some simple example computer code in several chapters to provide concrete ideas about how it works.

1.7 What Should Your Learning Objectives Be?

Here are some questions that we plan for you to be able to answer by reading and learning from this book. First, what are the important models developed in cognitive science for choice and decision-making? Second, what are the advantages of these models over traditional consumer or economic choice models? Third, when, or under what circumstances, should one cognitive model or another be used? Fourth, how does one estimate parameters and statistically test and compare these competing cognitive models? Fifth, how are these cognitive models related to neural structures and systems? If we succeed in answering some of these questions, then we have achieved our goal for writing this book.

2 Utility Theory

Utility theory provides a foundation for many of the subsequent choice models presented in this book. It has a long history dating back to the eighteenth century with early ideas developed by Daniel Bernoulli and Gabriel Cramer, and it continues to evolve to this day. The modern idea is that if a decision-maker obeys certain rational choice principles, then each option from a choice set can be represented by a utility, which is calculated using a formula that combines the various attributes of an option into a real number, and the decision-maker is predicted to choose the option producing the maximum utility. There is no need to assume that the decision-maker consciously carries out the steps in the utility formula. There is not even any reason to assume that utility causes the choice to be made. It is only necessary to assume that the decision-maker obeys the behavioral principles upon which the formula is derived, and then the formula can be used to predict the decision-maker's choice. Therefore, utility theories are not concerned with cognitive or neural processes; they are concerned only with the relation between the characterization of the options and the behavioral choices made by the decision-maker from these options.

2.1 Categories of Choice Situations

Before presenting any specific utility theory, we need to first identify and categorize different types of choice situations. Different types of utility theories are designed for different categories of choice situations. Table 2.1 presents six categories formed by two facets that are used to characterize the outcomes produced by choosing each option: the row represents the number of attributes used to describe each outcome, and the column represents

Table 2.1
Six ways to categorize choice situations.

	Certain	Risky	Uncertain
Single	No conflict	Playing the state lottery	Investing in a start-up company
Multiple	Buying a new car	Playing a game show to win prizes	Subjecting yourself to a new, untested medical procedure

the degree of uncertainty about the outcome. It also presents an example to illustrate each category.

The first row represents choices among options that are described by a single attribute, such as monetary return on investment for a business decision; alternatively, the single attribute could be nonmonetary, such as the number of lives saved by a medical decision. Of course, it is unrealistic to think that a choice involves only a single attribute; most choices involve many attributes. However, for practical purposes, it is often the case that one attribute is dominant and other attributes are less important or common across options, and so it is useful to model the choice using a single attribute. The second row represents more realistic cases in which the decision-maker considers multiple attributes, such as both the economic cost of a medical program and the number of lives saved by a medical program.

The first column represents choices made under certainty, in which we know the precise outcome produced by choosing each option. For example, if you buy a new car, you know exactly all the features and details about it. If there is only a single attribute (e.g., choosing between getting $10 or getting $11 for sure), then the decision is trivial and there is no conflict. However, if there are multiple attributes, then a conflict usually arises because one has to make trade-offs between attributes (e.g., trading quality for cost). Of course, it is unrealistic to think that the outcomes of any choice are completely certain (the new car may have issues, which is why it comes with a warranty and there also are lemon laws to protect the buyer). But for many practical scenarios, such as consumer purchases, it useful to model decisions in this way.

The second column represents choices made under risk. In this case, if a choice is made, different outcomes may be obtained. However, it is assumed that the probability of the outcome is known and agreed upon by all. In this sense, it is objective. For example, a choice scenario might involve a simple lottery whose outcome is determined by a coin flip (which everyone agrees is fair): perhaps the lottery has a .50 probability of producing heads, in which case a win of \$200 is obtained, and a .50 probability of tails, in which case a loss of \$100 is obtained. Once again, it is unrealistic to assume that any choice has precisely known and universally agreed-upon probabilities, but this assumption is a practical and useful approximation in some cases.

The third column represents decisions made under uncertainty. This is the more usual case, in which each option or course of action produces several possible outcomes, but there are no well-known or agreed-upon probabilities for these outcomes. The decision-maker may have only vague and uncertain information about the probabilities of the outcomes, and different decision-makers can disagree about these probabilities. For example, investing in a start-up company involves a lot of uncertainty about the possible return on investment, and a new medical procedure entails a lot of uncertainty about its positive and negative effects. It is often the case that uncertainty can be reduced by collecting information about the uncertain outcomes. For example, one can find and/or purchase new information (e.g., a drug company can conduct medical trials for a new medical procedure) or one can learn from experience (e.g., investing in a new product from a company with a past track record).

2.2 Weighted Additive Value Model

Let us first consider a model for the multiple-attribute and certain-outcome category, called the *weighted additive value model*. This model is used extensively in choice modeling. Suppose that each option, such as option X, is described by N attributes, and it has values $\{v_1, v_2, \ldots, v_N\}$ assigned to these attributes. Then the value assigned to option X equals

$$u(X) = w_1 v_1 + w_2 v_2 + \cdots + w_N v_N = \sum_j w_j v_j \,.$$

The decision-maker chooses the option that has the greatest weighted average value. The attribute values, v_j, $j = 1, \ldots, N$ are often considered to be a nonlinear transformation of the objective measurements or physical values of the attributes for an option. We will discuss this point later in this chapter, but for the time being, we will assume that these values are linearly related to objective value.

Table 2.2
Used car purchasing decision.*

Option	Mileage	Miles/Gallon	Price	Utility
	$w_M = -1/6$	$w_G = ½$	$w_p = -1$	Weighted sum
Car A	100K miles	25	\$15K	–\$19.17K
Car B	120K miles	22	\$14K	–\$23.0K
Car C	135K miles	22	\$12K	–\$23.5
Car D	90K miles	23	\$16K	–\$19.5
Car E	120K miles	21	\$14K	–\$23.5

What do the weights w_j represent? This is an important question, which is answered next. Later, we will show how the weights are determined.

Weights as exchange rates To understand the purpose of the weights, consider the example shown in table 2.2. Suppose that you are making a choice from a number of used cars. Each row describes one of these cars in terms of three attributes: mileage (i.e., the number of miles on the used car), miles per gallon (an accurate average estimated by the previous user) and the price in dollars. To keep the problem simple, suppose that these cars are all the same make and same year, and the cars are pretty much equivalent in all other attributes (otherwise, we would need to include them in a bigger table).

The weighted additive value model combines the values of the three attributes by adding the values within a row. However, the problem is that the units are not comparable across the attributes. You need to change the units of each attribute into a common currency before adding them.

You could choose any column for the common value scale, but the most natural column is Price. Therefore, you need to determine the exchange rate for miles per gallon and for mileage into dollars. That is, how much money are you willing to pay for a unit change in miles per gallon, and similarly for mileage? Once we know the exchange rate for an attribute, you can multiply this exchange rate by the attribute value of an option to produce the utility for that attribute and option.

The exchange rate for each attribute may depend on the person. One way to determine a person's exchange rate is to pose the following simpler problem, shown in table 2.3. Here, we have a choice between two options, and we assume that they are equivalent in all attributes except the two

Table 2.3
Figuring the exchange rate for mileage.

	Mileage	Price
Car Y	100K	\$15K
Car X	130K	\$*x*K

shown in the table. You can ask yourself—what value must \$*x*K be such that I would be indifferent between Car X and Car Y? Suppose that the answer is \$20K. Then you can determine the exchange rate as follows (assuming that our preferences are based on the additive utility rule):

$$u(Y) = (-100K \text{ miles}) \cdot w_M + (-\$15K)$$
$$u(X) = (-130K \text{ miles}) \cdot w_m + (-\$20K)$$

(setting $\$xK = \$20K$ to make us indifferent).

On the assumption that you are indifferent between these two when $\$xK = \$20K$, you solve for w_M as follows:

$$(-100K \text{ miles}) \cdot w_M + (-\$15K) = (-130K \text{ miles}) \cdot w_M + (-\$20K)$$
$$w_M = \frac{(-\$5K)}{30K \text{ miles}} = \frac{-1}{6}\frac{\$K}{K \text{ miles}}.$$

You also need the exchange rate for miles per gallon. Suppose that you used a similar procedure and found that $w_G = ½$ (*K* dollars per mpg). Then, for instance, you can compute the value for option A using the weighted added formula

$$u(A) = \frac{1}{6}\frac{\$K}{K \text{ miles}}(-100K \text{ miles}) + \frac{1}{2}\frac{\$K}{mpg}(25\, mpg) - \$15K = -\$19.17K.$$

More generally, you can compute the weighted additive value for each option using the formula

$$u(X) = w_M \cdot v_M(X) + w_G \cdot v_G(X) + w_P \cdot v_P(X),$$

where $v_M(X)$ is the value of Car X on attribute *M* (mileage), $v_G(X)$ is the value of Car X on attribute *G* (mpg), w_M is the exchange rate for mileage, w_G is the exchange rate for miles per gallon, and $w_P = -1$ by default. The last column of table 2.2 shows the utility for each option in dollars. If you used this model to guide your preference, then you should choose Car A, which has the maximum value (i.e., the least negative, in this case of cost).

One could argue that the exchange rate is not really constant across the entire range of values for an attribute. This objection could arise from the fact that the model, so far, has relied on a linear relation between the physical measurements (e.g., physical mileage of a car) and the subjective value of an attribute, described next.

Subjective values The model described here uses physical measurements (e.g., mileage) to represent the values entering the weighted additive model. This might fail to represent a person's preference for the following reason: for large changes in a physical value, it is usually necessary to include a subjective value function that maps that value into a subjective value. For example, most people perceive the change in subjective value for mileage produced by an increase from zero (brand-new car) to 10*K* miles as a larger change than the change in subjective value for mileage produced by an increase from 100*K* to 110*K* miles. The physical change is the same, but the subjective change is different. Furthermore, people may differ in their evaluation of the same objective or physical measurements of an attribute. For example, one person might think a change from 0 to 10*K* is a very large drop in the value of a car, but another person might consider this same change to be insignificant.

Methods for the simultaneous measurement of all the subjective values of all the attributes are available, but we will not go into the details of this more complicated measurement problem. Instead, we refer readers to more authoritative sources (e.g., Keeney & Raiffa, 1993). Instead, in many practical applications, choice modelers approximate the subjective value of a physically measured attribute (such as mileage) by a parametric function. For example, we could represent the subjective value of x mileage on a car by a power function $v(x) = \beta - (x^{\alpha})$ for $x \geq 0$ (see figure 2.1 for an example).

In summary, to apply the weighted additive model, we need to determine both the subjective value function $v_j(x)$ and the weight w_j for each attribute. Generally, the following procedure is used to do this: We estimate the coefficients (e.g., α) of the subjective value function, as well as the weights (e.g., w_j) of the weighted additive model by fitting the entire model to a data set of choices obtained from an individual using a type of nonlinear regression analysis called a *multinomial logit model*. We show how this is done in chapter 3.

Additive model axioms The weighted additive value model assumes that the decision-maker obeys a small number of necessary axioms (Krantz,

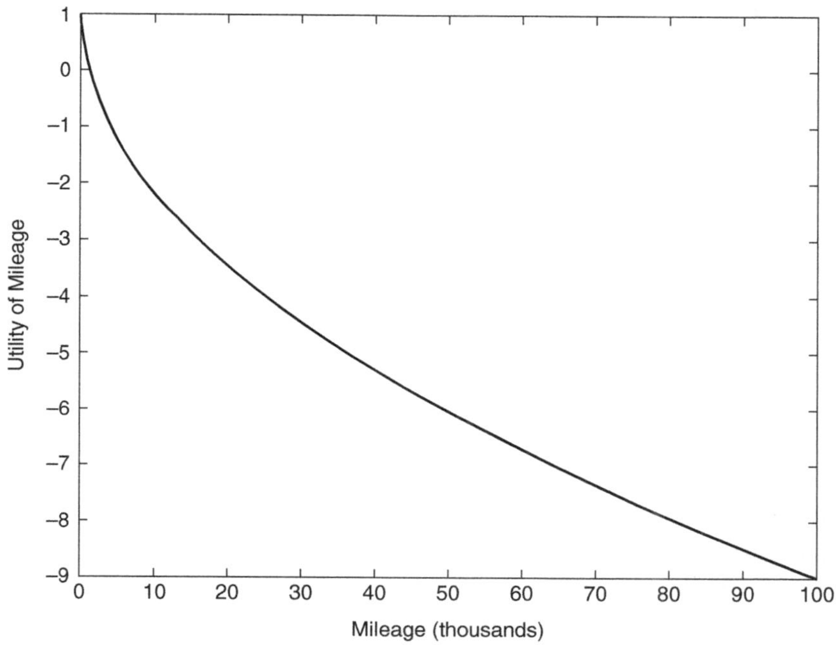

Figure 2.1
Subjective value $v(x)$ of mileage (x=physical mileage in thousands of miles) determined by $v(x)=1-(x^5)$. Other parametric value functions are possible as well, such as a logarithmic function $v(x)=\beta-\log(\alpha+x)$.

Luce, Suppes, Tversky, 1971, Krantz & Tversky, 1972). One of these axioms is transitivity. For example, if a person prefers Car A over Car B and Car B over Car C, then the person should prefer Car A over Car C. This is necessary because the additive value model assigns a real number to each option, and real numbers are transitive. Transitivity is often justified as an axiom of choice behavior by arguing that violations of this axiom permit a person to be turned into a money pump. Suppose that a person starts out by owning *C*. If a person prefers *B* to *C*, then the person should be willing to pay money to exchange *C* for *B* and end up with *B* and a loss from the exchange. Likewise, if the person prefers *A* to *B*, then the person should be willing to pay money to exchange *B* for *A* and end up with *A* and a loss from the second exchange. Finally, if the person also prefers *C* to *A*, then the person should be willing to pay to exchange *A* for *C*, and thus end up returning to the original position *C*, but after losing money on three

Table 2.4
Design for testing joint independence.

Car 111	Car 121	Car 211	Car 221
Low price	Low price	High price	High price
Low mileage	High mileage	Low mileage	High mileage
Low mpg	Low mpg	Low mpg	Low mpg
Car 112	**Car 122**	**Car 212**	**Car 222**
Low price	Low price	High price	High price
Low mileage	High mileage	Low mileage	High mileage
High mpg	High mpg	High mpg	High mpg

exchanges. However, it is unlikely that these intransitive preferences could be exploited by building a money pump. Instead, people will presumably notice their intransitive cycles at some point, making the money pump a bogeyman that only demonstrates in principle the irrationality of intransitive choices, but one that is not observed in practice.

Another axiom is joint independence, which can be illustrated with the help of table 2.4. This table illustrates a 2 (price) $\times$ 2 (mileage) $\times$ 2 (mpg) factorial design with eight cells representing eight car options. A car is labeled Car *i, j, k*, where $i = $ L, H represents the low versus high level of the price, $j = $ L, H represents the low versus high level of the mileage, and $k = $ L, H represents the low versus high level of the miles per gallon. According to the joint independence axiom, if we prefer the four cars 111,112,121,122 in a particular order when price is low, then we should observe the same order of the four cars 211, 212, 221, 222 when price is high. More generally, if we change the labeling of cars in the tables (e.g., relabel *i* as mileage, *j* as mpg, and *k* as price), joint independence requires the ordering of two factors, holding a third fixed, to be consistent across different fixed values of the third factor.

It is relatively easy to see why the joint independence axiom is required by additive models. Our most general additive model can be written as

$$u(Car_{ijk}) = w_M \cdot v_{Mi} + w_G \cdot v_{Gj} + w_P \cdot v_{Pk},$$

where $v_{M,i}$ is the value of the *i*th value of attribute *M*, $v_{G,j}$ is the value of the *j*th value of attribute G, and $v_{P,k}$ is the value of the *k*th value of attribute P. If we find that $\text{Car}_{1,j,k}$ is preferred to $\text{Car}_{1,j'k'}$, then this implies the following:

$$\begin{aligned} u(Car_{1,jk}) &= w_P \cdot v_{P,1} + w_M \cdot v_{M,j} + w_G \cdot v_{G,k} \\ &> u(Car_{1,j'k'}) = w_P \cdot v_{P,1} + w_M \cdot v_{M,j'} + w_G \cdot v_{G,k'} \\ &\rightarrow w_M \cdot v_{M,j} + w_G \cdot v_{G,k} > w_M \cdot v_{M,j'} + w_G \cdot v_{G,k'} \\ &\rightarrow u(Car_{2,j,k}) = w_P \cdot v_{P,2} + w_M \cdot v_{M,j} + w_G \cdot v_{G,k} \\ &> u(Car_{2,j'k'}) = w_P \cdot v_{P,2} + w_M \cdot v_{M,j'} + w_G \cdot v_{G,k'}, \end{aligned}$$

where the last inequality implies the satisfaction of joint independence.

The way that a decision theorist views an additive model of preference is as follows: If a person satisfies the axioms of transitivity and joint independence (as well as another technical axiom related to solvability, which we do not present here) for all preferences from a set of choice options, then we can predict this person's preferences using a weighted additive utility model. This does not mean that the person actually computes a weighted additive combination in their head. It only means that we can predict their behavior using this rule, whatever process they use to make their choices.

2.3 Averaging Model

Another model that can be used for the multiple-attribute and certain-outcome category is the *weighted averaging model* (Anderson, 2014). The weighted averaging model seems very similar to the weighted additive model, but it differs in several important ways. Suppose that a personnel manager is choosing among applicants for a job, and the applicants vary according to test scores (T), experience (E), and recommendation letters (R). A weighted averaging model would recommend choosing the candidate with the largest weighted average value:

$$u(Candidate_{ijk}) = \frac{w_T}{w_T + w_E + w_R} \cdot v_{Ti} + \frac{w_E}{w_T + w_E + w_R} \cdot v_{Ej} + \frac{w_R}{w_T + w_E + w_R} \cdot v_{Rk},$$

where the weights are nonnegative. This model is useful when the set size, or number of attributes, varies across the choice options. For example, one candidate may come from a situation in which the test scores are not applicable, and so in this case, the weighted average value assigned to this candidate is the average:

$$u(Candidate_{jk}) = \frac{w_E}{w_E + w_R} \cdot v_{Ej} + \frac{w_R}{w_E + w_R} \cdot v_{Rk}.$$

Note that the relative weight assigned to an attribute such as experience changes depending on whether the test scores are available. This change in relative weight with set size produces a nonadditive effect, as illustrated in figure 2.2. If we vary the experience from low to high, the values changes in the same way, from v_{EL} to v_{EH} for both set sizes. But the weight is greater for the set size 2 case, and so the same change has a larger effect, which produces a crossover interaction.

The weighted averaging model makes another interesting prediction that differs from the weighted additive model. Suppose that an applicant has a high test score (e.g., $v_T = 9$), good experience (e.g., $v_E = 9$), and a positive recommendation letter (e.g., $v_R = 7$). Should this candidate submit the recommendation along with the test scores and experience description? Assuming equal weights (=1), according to a weighted additive model, the positive value of the recommendation would increase the total value

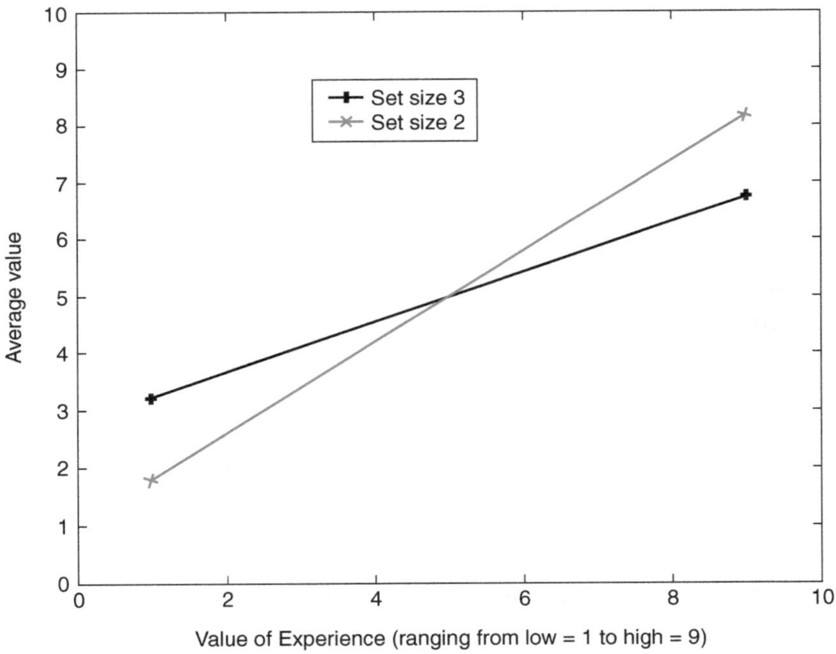

Figure 2.2
Crossover interaction effect predicted by the weighted averaging model when the set size is changed.

(9+9+7=25) compared to not adding it to the total value (9+9=18); but according to a weighted averaging model, averaging in the recommendation would lower the average value (25/3=8.33) compared to not including the recommendation (18/2=9). This kind of evidence for a crossover interaction supporting a weighted averaging rule (as opposed to a weighted additive rule) was reported by Troutman and Shanteau (1976).

2.4 Intertemporal Choice and the Multiplicative Model

In the previous example, the weighted additive model was applied to a consumer choice type of problem. There are other applications, and an important one is the intertemporal problem (Loewenstein & Elster, 1992). Suppose that a person is facing a choice among several different consumption streams. A *consumption stream* is a sequence of certain outcomes delivered at different points in time $X=\{x_{t1}, x_{t2}, \ldots x_{tN}\}$, where t_j, $j=1 \ldots N$ is the time index. The *delay discounting model* is a weighted additive model that weights the value of each outcome by a factor that generally decreases as the time increases:

$$u(X\ stream) = \sum_{j=1}^{N} w(t_j) \cdot v(x_j).$$

Economic models often assume an exponentially decreasing weight $w_t=e^{-\alpha t}$, $\alpha>0$, but psychologists have found empirical evidence for what is called *hyperbolic discounting* $w_t = \frac{1}{1+\alpha t}$, $\alpha > 0$ (Loewenstein & Elster, 1992). The reason for choosing the hyperbolic model over the exponential model is that the exponential model requires time consistency, which people often violate. The hyperbolic model predicts violations of time consistency. For example, suppose a person prefers a small immediate gain of x\$ over a larger gain of y\$ that is delayed until time t_1 months later. Time consistency requires that if we increase the delivery times for both options by the same amount of t_2 months, then this person should continue preferring x\$ with a delay of t_2 months over the larger gain of y\$ that is delayed until t_1+t_2 months. However, people often violate this time consistency property by reversing their preferences with the shifted time points, which turns out to be predicted by the hyperbolic model.

Multiplicative model Researchers have frequently investigated the temporal discounting model at a single time point. For example, a person may be asked to choose between an immediate small payoff and a larger delayed payoff. When the model involves only one time point, it reduces to a multiplicative model: the utility of obtaining payoff x at time t equals $u(x, t) = w_t u(x)$. If the payoffs are restricted to be positive, then this is equivalent to an additive model on a log scale $\ln(u(x, t)) = \ln(w_t) + \ln(u(x))$. However, if the payoffs can change sign from positive to negative, then this log transformation is no longer possible. In the latter case, preferences can be sign dependent: A person's preference for a delayed outcome decreases with the delay when the outcome is positive (e.g., receiving a positive return on an investment), but the person's preference for a delayed outcome increases with the delay when the outcome is negative (e.g., paying a return on a loan). Therefore, the effect of the delay changes direction depending on the sign of the outcome to produce a crossover type of interaction effect. If the preferences are sign dependent, then we must use the multiplicative model.

Another important application of the multiplicative model is to the quality and duration of life (Miyamoto & Eraker, 1988). The recommendation is to choose the larger product of quality and life duration: $u(Q, D) = v_Q(Q) \cdot v_D(D)$. If duration of life is positive, then increasing quality increases the overall value. But if duration of life is zero, then increasing quality has no effect at all. Alternatively, if quality of life is positive and above some tolerable level, then increasing duration increases the total value. But if quality of life is below some tolerable level, then increasing duration could decrease the value. This change in direction of the effect of delay with sign of quality produces a crossover interaction effect. In sum, the preferences for quality and duration are sign dependent, so a multiplicative model is needed.

2.5 Expected Utility Theory

Next, we consider models for the single-attribute and risky-outcome category. The first is called the *expected value model* and the second is the *expected utility model*. Suppose that $\{x_1, x_2 \ldots., x_N\}$ is a set of possible single attribute outcomes (e.g., possible monetary returns on an investment). A risky option X produces outcome x_1 with probability p_1, and outcome x_2 with probability p_2, and so on, up to outcome x_N with probability p_N. If x_j is the objective dollar value of investment j, then the expected value of option X equals

$$EV(\textit{risky option } X) = \sum_{j=1}^{N} p_j \cdot x_j.$$

Table 2.5
Risky choice between two medical treatments.

	State A (.90 probability)	State B (.10 probability)
Surgery	0 years to heal	50 years to heal
Physical therapy	7 years to heal	5 years to heal

Intuitively, the expected value represents the long-run average payoff obtained by playing the gamble over and over again. At first it seems reasonable to choose the risky option that produces the highest expected value. However, let's take a closer look at this issue.

Table 2.5 describes a choice between two medical treatments for a knee injury: surgery and physical therapy (shown as the rows). The outcomes for each option depend on the state of the knee. If state A is present, then the knee heals back to normal in weeks (rounded off to zero years), but the physical therapy would take 7 years to return to normal use. However, if state B is present, surgery will almost permanently damage the knee (truncated at 50 years to heal in the table), but physical therapy would take only 5 years to get back to normal. Assume that the decision-maker knows the probability of each state, perhaps based on medical base rates for cases very similar to this one. Then the expected value for each option equals (years waiting is treated as a negative value)

$$EV(\text{surgery}) = (.90)(0 \text{ years}) + (.10)(-50 \text{ years}) = -5.0 \text{ years}$$

$$EV(\text{therapy}) = (.90)(-7 \text{ years}) + (.10)(-5 \text{ years}) = -6.8 \text{ years}.$$

According to expected value, surgery should be chosen because it produces the maximum expected value. However, this choice seems rather risky, and a person may not find this answer to be satisfactory.

The problem with using expected value for predicting preferences is that it relies on physical or objective measurements (e.g., years to heal). This assumes that the change in value from 0 years to 10 years equals the change in value from 40 years to 50 years. However, many people might value the change from 0 years to 10 years more than the change from 40 years to 50 years of waiting to heal. In other words, we need to replace the physical measurement of years to wait with the subjective utility of waiting. Consequently, a person needs to assign utilities $u(x)$ to the years $x = 0$, 5, 7, and 50. These utilities will be personal and vary across individuals.

More generally, rather than maximizing expected value, a person maximizes expected utility, defined as follows:

$$EU(X) = \sum_j p_j \cdot u(x_j),$$

where the utility $u(x_i)$ of an outcome replaces the objective value of the outcome. But how can we measure a person's utility function, $u(x_i)$?

First, the utility function is only an interval level scale, which means that the location and unit of the scale is arbitrary. This follows from the fact that we can add a constant to all the utilities and we can multiply all the utilities by a positive constant without changing the order of the expected utilities. Thus, we can set two utilities to arbitrary values and choose two convenient values. We set the worst outcome to zero and the best outcome to 1. In this example, we set $u(50 \text{ years}) = 0$ and $u(0 \text{ years}) = 1$. However, we still need to determine the utilities for the intermediate values, 5 and 7 years.

To determine the utility of 7 years, we present the decision-maker with the simple choice shown in table 2.6, which presents two risky options. Option X produces the best outcome (0 years) with probability p and the worst outcome (50 years) with probability $(1-p)$. Option Y produces the outcome $x = 7$ years for sure—this is the outcome for which we wish to determine the utility. In table 2.6, we need to determine the probability p that makes the decision-maker indifferent between gambles X and Y. Suppose that a person is indifferent with the probability set to $p = .95$, so that in this case, the person is indifferent between X and Y. Then we can find the utility of $x = 7$ as follows:

$$\begin{aligned} EU(X) &= p \cdot u(0) + (1 - p) \cdot u(50) \\ &= (.95) \cdot 1 + (.05) \cdot 0 = .95 \\ &\text{(assuming indifference)} \\ EU(Y) &= u(7). \end{aligned}$$

In this way, we have solved for the utility of 7 years and found it to be $u(7) = .95$ for this person. A similar procedure can be used to determine the

Table 2.6
Probability equivalence procedure for determining a utility of an outcome x.

	Probability p	Probability $1-p$
Option X	0 years	50 years
Option Y	$x = 7$ years	$x = 7$ years

utility of 5 years, and suppose that the utility of 5 years turns out to be $u(5) = .98$. Then we can return to the original problem and compute the expected utility of surgery versus therapy:

$$EV(\text{surgery}) = (.90)(1) + (.10)(0) = .90 \text{ utiles}$$

$$EV(\text{therapy}) = (.90)(.95) + (.10)(.98) = .953 \text{ utiles.}$$

Now the maximum expected utility is the therapy treatment for the decision-maker. In other words, using the utility function, the decision-maker has become more conservative, or less risky.

In practice, researchers often use a simple function to approximate a person's utility function. For example, money is often assumed to have a decreasing marginal utility. In other words, the increase in wealth from zero (bankrupt) to $1 million seems to provide much greater subjective utility than the increase in wealth from $100 million to $101 million. To capture this decreasing marginal utility of money, a concave utility function is often used, such as a power function $u(x) = x^{\alpha}$ for $x \geq 0$ and $0 \leq \alpha \leq 1$ (see figure 2.3). A risk averse decision-maker will prefer a sure value x over a gamble X that has an expected value equal to $EV(X) = x$, and concave utility functions produce risk averse preferences. Risk aversion and concave utility functions are used to explain why people buy insurance. The expected value of an insurance policy is negative for the purchaser (and positive for the insurance company).

Axioms of expected utility theory Why should a person follow the expected utility rule? Is this the rational way to make decisions? One might try to argue that this rule is rational because it maximizes the long-run average utility. However, this argument does not make much sense for a personal decision that is made only once, like the knee surgery problem shown in table 2.5. A more basic justification is needed in order to claim that this rule is rational.

The rational basis for the expected utility rule relies solely on the acceptance of a small set of intuitively compelling, rational, axioms. If a decision-maker obeys these axioms, then this person's preferences can be predicted using the expected utility rule. The original derivation of the expected utility rule from axioms was presented by Von Neumann and Morgenstern (2007/1945). Later, other derivations appeared, and there turns out to be more than one way to formulate the axioms, but they are all very similar. Next, we present one formulation by Luce and Raiffa (1957) and show how the expected utility formula can be derived from these axioms.

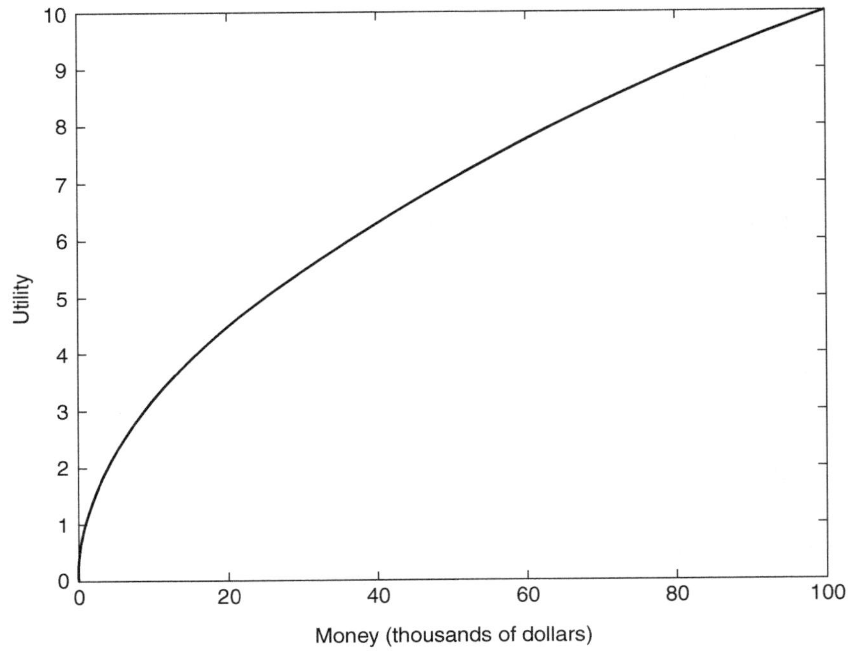

Figure 2.3
The utility of money determined by a utility function that is strictly concave produces risk aversion.

First, we suppose that there is a finite set of real outcomes $S = \{x_1, x_2 \ldots., x_N\}$, ordered such that $x_1 \leq_P x_2 \ldots \leq_P x_N$, where $x_i \leq_P x_j$, symbolizes that the relation x_j is preferred or indifferent to x_i. We define a gamble such as X as the probability distribution over these outcomes $X = \{p_1$ to get x_1, p_2 to get x_2, ..., p_N to get $x_N\}$, where $p_i = p(X = x_i)$.

The first axiom is transitivity: if gamble X is preferred or equal to Y and Y is preferred or equal to Z, then X is preferred or equal to Z.

The second axiom is dominance: if outcome x_L is preferred or equal to x_S, and $p_L \geq p_S$, then gamble X, *which* obtains x_L with probability p_L and otherwise x_S, is preferred to gamble Y, which obtains x_L with probability p_S and otherwise x_S.

The third axiom is substitutability: if gamble U is indifferent to the real-value payoff x, then we can substitute gamble U for the real value x.

The fourth axiom is solvability: if x_L is preferred to x and x is preferred to x_S, then we can find a probability p such that we are indifferent between x and the gamble that obtains x_L with probability p and otherwise x_S.

The final axiom is reduction: we can use standard probability theory to reduce any probability mixture of gambles to a gamble described solely in terms of the final outcomes in S.

The Expected Utility theorem states that if a person agrees to all these axioms, then we can predict their preferences using the expected utility rule.

Now let us see how the expected utility rule can be derived from these axioms. Suppose that we face a choice between two risky options, X and Y.

By the reduction axiom, we can always reduce them to probability distributions over the final outcomes: $X = \{p_1, p_2, \ldots, p_N\}$ and $Y = \{q_1, q_2, \ldots, q_N\}$. Suppose that the worst outcome is x_1 and the best outcome is x_N. Define a gamble U_i as a probability u_i to obtain x_N and otherwise obtain x_1.

By the solvability axiom, we can find a probability u_i such that we are indifferent between gamble U_i and outcome x_i.

By the substitutability axiom, we can substitute gamble U_i for the outcome x_i within each gamble X and Y. After making this substitution, we can express the gambles as $X = \{p_1$ to get U_1, p_2 to get U_2, ..., p_N to get $U_N\}$ and $Y = \{q_1$ to get U_1, q_2 to get U_2, ..., q_N to get $U_N\}$. After substitution, each gamble X and Y has only two final outcomes, x_1 and x_N, because each gamble U_i is a probability u_i to obtain x_N otherwise x_1.

Applying reduction once more to the two final outcomes, we can define

$$EU(X) = (p_1 \cdot u_1 + p_2 \cdot u_2 + \cdots + p_N \cdot u_N)$$

as the probability of winning the best outcome x_N for gamble X, and

$$EU(Y) = (q_1 \cdot u_1 + q_2 \cdot u_2 + \cdots + q_N \cdot u_N)$$

as the probability of winning the best outcome x_N for gamble Y. Note that $EU(X)$ and $EU(Y)$ are both expressed as probabilities now. This produces the equivalent expression for the two gambles:

$X = \{1 - EU(X)$ to get x_1, $EU(X)$ to get $x_N\}$ and

$Y = \{1 - EU(Y)$ to get x_1, $EU(Y)$ to get $x_N\}$.

According to dominance, if $EU(X) > EU(Y)$, then we prefer gamble X over Y; if $EU(Y) > EU(X)$, then we must choose gamble Y over X; and if $EU(X) = EU(Y)$, then we are indifferent.

By setting the utility of outcome $u(x_i)$ equal to probability u_i in gamble U_i, we thus arrive at the expected utility rule from axioms.

Empirical tests of the EU rule The expected utility rule is a fairly general rule. After all, it allows decision-makers to have their own personal utility function. If two people disagree about their preferences concerning a single pair of options, it is not possible to say that one person is correct or more rational. Instead, they simply have different subjective utilities. However, it is possible to claim that a person is irrational if (1) the person believes that the axioms of expected utility theory are rational, and (2) the person violates one or more of those axioms. Do people violate the axioms, then?

As we mentioned earlier, there are different but closely related ways to formulate the axioms of expected utility theory. Another formulation uses the axiom of independence instead of substitutability to obtain the same expected utility rule. The independence axiom of expected utility is closely related to the independence axiom of weighted additive value models. This follows from the fact that the expected utility model adds the contributions across the outcomes of a gamble. According to the independence axiom of expected utility theory, if a person prefers gamble X over gamble Y, then the person should prefer gamble $X' = \{p_Z$ to get gamble X, $(1 - p_Z)$ to get gamble $Z\}$ over gamble $Y' = \{p_Z$ to get gamble Y, $(1 - p_Z)$ to get gamble $Z\}$. This is a necessary axiom for the expected utility rule because the common gamble Z cancels out when comparing the expected utilities:

$$\text{if } EU(X) > EU(Y), \text{ then}$$
$$EU(X') = p_Z \cdot EU(X) + (1 - p_Z) \cdot EU(Z) >$$
$$EU(Y') = p_Z \cdot EU(Y) + (1 - p_Z) \cdot EU(Z).$$

Beginning with a seminal article by Allais (1990), the independence axiom has been empirically tested and found to be violated in numerous studies (Kahneman & Tversky, 1979). (One important issue is that these results are based on aggregation across individuals rather than tests at the individual level.)

Consider the following simple example. Suppose that you are given a choice between

X, which gives $3,000 for sure

and

Y, which obtains $4,000 with probability .80, and otherwise nothing.

Although the expected value of gamble *Y* is considerably greater than the sure thing *X*, most people are risk averse and prefer the sure thing *X*. No axiom is violated by this preference. However, now consider a choice between

X′, which provides .25 chance of winning $3,000, or else nothing,

and

Y′, which has a .20 chance of winning $4,000, or else nothing.

Most people prefer gamble *Y′*. But this violates the independence axiom (actually the reduction plus independence axiom). To see that this is a violation, note that gamble *X′* can be reexpressed (using the reduction axiom) as a probability mixture of winning *X* with probability .25 or zero with probability .75. Likewise, gamble *Y′* can be reexpressed (using the reduction axiom) as a probability mixture of winning gamble *Y* with probability .25 or zero with probability .75. Therefore, the independence axiom applies, so if *X* is preferred to *Y*, then *X′* should be preferred to *Y′*. Another way to see this is expressed as follows: if $EU(X) > EU(Y)$, then $EU(X') = .25 \cdot EU(X) + .75 \cdot (0) > EU(Y') = .25 \cdot EU(Y) + .75 \cdot (0)$. The fact that a majority of people violate independence has caused some theorists to question its status as a rational axiom.

Prospect theory In view of empirical violations of expected utility theory, alternative models were proposed that were perhaps less rational, but more descriptive of actual human behavior. The most popular was *prospect theory*, which was proposed by Kahneman and Tversky (1979). We are skipping some details about prospect theory in this discussion; we will present only the "stripped down" version of the theory, which contains the essential ideas. Prospect theory was originally restricted to gambles with two or fewer non-zero outcome gambles, which were called *prospects*. An example of a prospect is prospect *X*, which gives a p_W probability to obtain *x*, otherwise obtain *y*.

First, prospect theory proposes that people evaluate the outcomes x, y relative to a reference point, denoted as *r*. If an outcome *x* is above the reference point (e,g., $x-r>0$), then it is a gain; if an outcome *y* is below the reference point (e.g., $y-r<0$), then it is a loss. In addition to risk aversion, prospect theory assumes that people exhibit loss aversion (i.e., losses loom

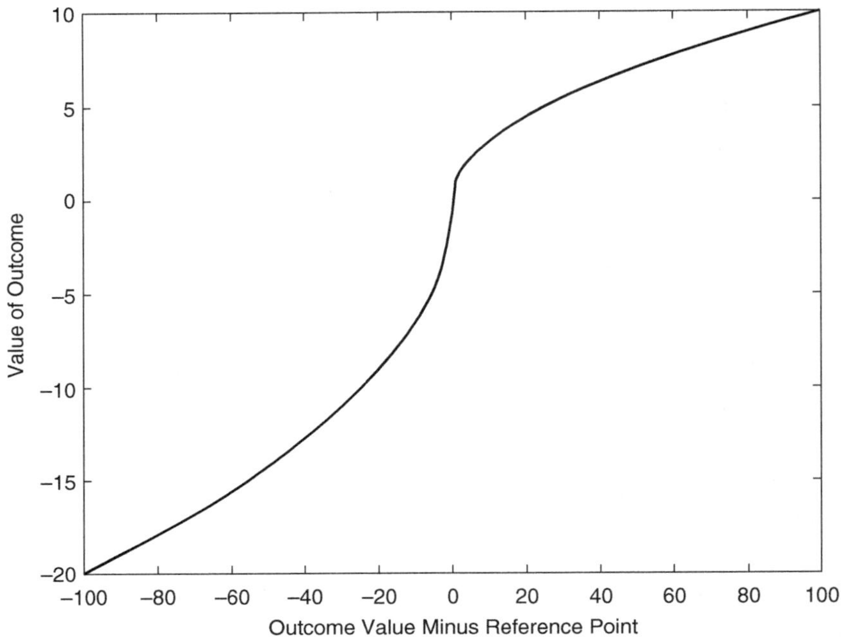

Figure 2.4
Prospect theory value function using $\alpha = .50$ and $\lambda = 2$.

larger than gains), so that the effect of an objective change in loss is larger than the effect of the objective change in gain of the same size. In many applications, the value function of prospect theory is approximated by a power function of the form

$$v(x) = x^{\alpha} \text{ for gains, and } v(x) = -\lambda \cdot |x|^{\alpha} \text{ for losses.}$$

Parameter α determines the risk aversion, and parameter λ determines the loss aversion (see figure 2.4 for an example).

Second, prospect theory assumes that probabilities are interpreted in a subjective manner. More formally, it is assumed that the objective probability p_i of outcome x_i is nonlinearly transformed into a decision weight, $w(p_i)$. In practice, a nonlinear function such as $w(p) = \frac{p^{\alpha}}{(p^{\alpha} + (1 - p^{\alpha}))}$ is often used to approximate the decision weight function (see figure 2.5 for an example) According to prospect theory, the decision-maker chooses the prospect that maximizes the weighted sum:

$$PU(X) = w(p_x) \cdot v(x) + w(p_y) \cdot v(y).$$

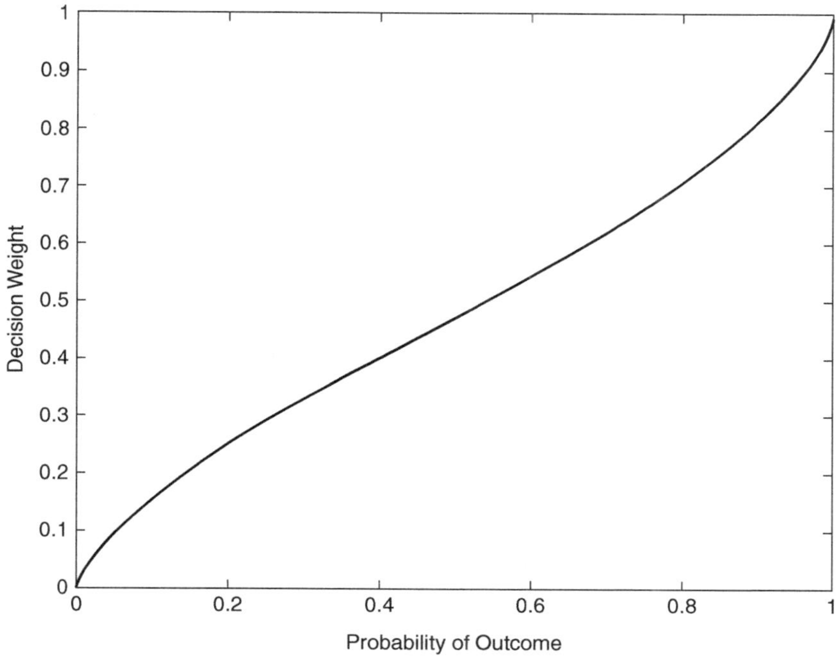

Figure 2.5
Decision weighting function using $\alpha = .75$.

How does prospect theory explain violations of independence? The nonlinear weighting function is critical for this account. Consider once again the pair of gambles:

X, which gives $3,000 for sure,

and

Y, which obtains $4,000 with probability .80, and otherwise nothing.

According to prospect theory, the fact that people prefer *X* to *Y* implies that

$w(1) \cdot v(3{,}000) > w(.80) \cdot v(4{,}000)$, which implies that

$v(3{,}000)/v(4{,}000) > w(.80)/w(1.0)$.

Next, consider once again the following pair of gambles:

X′, which provides .25 chance of winning $3,000, or else nothing,

and

Y′, which has a .20 chance of winning $4,000, or else nothing.

The fact that people now prefer Y' over X' now implies

$w(.25) \cdot v(3{,}000) < w(.20) \cdot v(4{,}000)$, which implies that

$w(.20)/w(.25) > v(3{,}000)/v(4{,}000)$.

Together, these statements imply

$w(.20)/w(.25) > v(3{,}000)/v(4{,}000) > w(.80)/w(1.0)$ which implies

$w(.25 \cdot .80)/w(.25 \cdot 1.0) > w(.80)/w(1.0)$.

In other words, the weighting function is inhomogeneous, so $w(a \cdot x) \neq a \cdot w(x)$.

Beyond prospect theory Soon after prospect theory was developed, challenges to it appeared. One of the most important challenges was the fact that prospect theory nonlinearly transformed the probabilities of outcomes. Nonlinear transformations of the outcome probabilities produce violations of the principle of dominance, which is very irrational. Suppose that the weighting function is nonlinear, so $w(p) + w(1-p) < 1$. For example, using the weighting function shown in figure 2.5, we find that $w(.50) = .4719$ so that $w(.50) + w(.50) = .9439$. If $w(p) + w(1-p) < 1$, then $w(1-p) < 1 - w(p)$, and so it follows that $[1 - w(p)]/w(1-p) > 1$. Suppose that we pick two outcomes x and $(x+y)$, where $y>0$ is sufficiently small so that

$[1 - w(p)]/w(1-p) > v(x+y)/v(x) > 1$.

Now consider a choice between the following two prospects:

X: win x for sure.

Y: win $(x+y)$ with probability p, and otherwise x.

Clearly Y dominates X because Y produces either the same as X or a slightly larger win than X, and no one chooses X. However, note that

$1 - w(p)/w(1-p) > v(x+y)/v(x)$ implies

$v(x) > w(p) \cdot v(\mathrm{x}) + w(1-p) \cdot v(x+y)$.

Thus, if we use the weighted sum of outcomes $w(p) \cdot v(\mathrm{x}) + w(1-p) \cdot v(x+y)$ to represent the prospect value of gamble Y, then this prospect value is below the value of the sure thing for X. This is clearly a bad prediction. Recall that here, we are using the stripped-down version of prospect theory. To avoid problems like this, prospect theory included a stage that edited out the guaranteed gain as follows: $PU(Y) = v(\mathrm{x}) + w(p) \cdot v(y)$.

In light of problems like those described here, the original 1979 prospect theory was later replaced by what is called *cumulative prospect theory*

(Tversky & Kahneman, 1992). Cumulative prospect theory builds on earlier work by psychologists (Lopes, 1987; Birnbaum & Stegner, 1979) and economists who developed what is called *rank dependent utility theory* (e.g., Quiggin, 1982). This new theory was needed for two reasons. One was to avoid the problems of violating dominance produced by the nonlinear transformation of the probabilities of the outcomes (mentioned previously). The second was to obtain a more satisfactory model that was not limited to two outcomes, but instead applied to the more general case of N outcomes.

We will not go into all the details concerning cumulative prospect theory, but only provide a brief description. The general theory applies to positive and negative outcomes, but for simplicity and brevity, we will restrict our attention to positive outcomes. Cumulative prospect theory assigns the following utility to each prospect:

$$CPU(X) = \sum_{x} w(x_i)\, u(x_i)$$

$$w(x_i) = \pi\left(\sum_{j=i}^{n} p_j\right) - \pi\left(\sum_{j=i+1}^{n} p_j\right).$$

The sum $\sum_{j=i}^{n} p_j$ is the probability of obtaining outcome x_i or better, the sum $\sum_{j=i+1}^{n} p_j$ is the probability of getting an outcome strictly better than x_i and π is an increasing nonlinear function.

Essentially, cumulative prospect theory changed the way the decision weights are computed: instead of applying the nonlinear transformation directly to the probability of outcomes, the nonlinear transformation is applied to the decumulative distribution of outcomes to produce a transformed decumulative distribution, and the decision weights are computed from the decumulative distributions. (The decumulative distribution is 1 minus the cumulative distribution.)

It is interesting to note that one general property that cumulative prospect theory (as well as all rank dependent utility theories) must satisfy is the property of stochastic dominance. Recall that a gamble is a probability distribution across the outcomes. One gamble X stochastically dominates another gamble Y if the decumulative distribution of X always equals or exceeds Y (and exceeds at least 1 point). Unfortunately, people do not seem to obey this property either. Recently, strong evidence for violations of stochastic dominance produced by event splitting manipulations have been reported (Birnbaum, 2008b), who has an alternative utility model that actually predicted these results called the Transfer of Attention Exchange (TAX) theory.

2.6 Subjective Expected Utility Theory

Now we consider a model for the single-attribute and uncertain-outcome category called the subjective expected utility model originally developed by Savage (1972/1954). Table 2.7 illustrates the basic concepts used in the subjective expected utility theory.

The table shows two rows representing two different actions that a decision-maker can take, and the column shows N different mutually exclusive and exhaustive possible uncertain states of the world. An action assigns an outcome to each state of the world; for example, x_i is the outcome produced by choosing action X and the state of world E_i obtains, and y_i is the outcome produced by choosing action Y and the state of world E_i obtains. The key idea is that there are no exactly known or objective probabilities for the states of the world, and each person must assign a personal probability, $s(E_i)$, to each state, such as E_i. The probabilities are subjective because different individuals may disagree about the probabilities assigned to each state.

Although the probabilities are subjective, subjective expected utility theory is a rational decision theory: it is based on axioms which guarantee that the subjective probabilities obey the standard rules of probability theory. In particular, the subjective probabilities range between zero and 1, with zero assigned to the impossible event, and 1 assigned to the certain event, and given that the states E_1 and E_2 are mutually exclusive, then the $s(E_1 \text{ or } E_2) = s(E_1) + s(E_2)$. If an individual obeys these and other rational axioms proposed by Savage, then this person should prefer (or be indifferent) for X over Y if

Table 2.7

Rows are actions and columns are states.*

	State E_1	State E_2	...	State E_N
Action X	x_1	x_2	...	x_N
Action Y	y_1	y_2	...	y_N

*Ellipses represent additional states

$$SEU(X) = \sum_j s(E_j)\, u(x_j) \geq \sum_j s(E_j)\, u(x_j) = SEU(Y).$$

Like the expected utility model, one of the axioms of subjective expected utility theory is the independence axiom. According to this axiom, if exactly the same payoff is assigned to a particular state of the world for both actions *X* and *Y*, then the preference for *X* versus *Y* should not depend on the value of the common payoff. This axiom was challenged using a decision problem designed by Ellsberg (1961), shown in table 2.8.

In this problem, the payoff is determined by a colored ball randomly drawn from an urn. There are 99 balls in the urn; 33 are known to be red, and the remaining 66 are either blue or green. However, the exact number of blue or green balls is unknown. The payoff for each action is determined by the ball selected. For example, if *X* is chosen and a red ball is selected, then the payoff is $100, and otherwise the payoff is zero.

According to subjective expected utility theory, if the green ball appears, then a common payoff (zero) is obtained from both *X* and *Y*. Similarly, if the green ball appears, then a common payoff ($100) is obtained from both *X′* and *Y′*. Therefore, the common value of this payoff should not matter whether it is zero or $100 because the common payoff produced by the green ball just cancels out. However, if we cancel out and ignore the green ball event, then the choice between *X* and *Y* is the same as the choice between *X′* and *Y′*. Therefore, according to the independence axiom, if a person prefers *X* over *Y*, then the person should also prefer *X′* over *Y′*.

When faced with a binary choice between action *X* versus action *Y*, most people prefer action *X* because they know that there is a 33/99 chance to win $100 and they don't know what the chance of winning is for action *Y*. However, when faced with a binary choice between action *X′* versus action *Y′*, now most people prefer *Y′* because they know that there is a 66/99

Table 2.8
Example decision under uncertainty.

Action	Red Ball	Blue Ball	Green Ball
X	$100	0	0
Y	0	$100	0
X′	$100	0	$100
Y′	0	$100	$100

chance of winning \$100 with Y', but now they don't know the chance of winning for X'. This preference pattern violates the independence axiom.

Another way to see that this preference pattern violates subjective expected utility theory is the following simple analysis. According to the theory, if X is preferred to Y, then

$SEU(X) = s(\text{Red}) \cdot u(\$100) > SEU(Y) = s(\text{Blue}) \cdot u(\$100)$, which implies

$s(\text{Red}) > s(\text{Blue})$.

But if Y' is preferred to X', then

$SEU(Y') = s(\text{Blue}) \cdot u(\$100) + s(\text{Green}) \cdot u(\$100) >$

$SEU(X') = s(\text{Red}) \cdot u(\$100) + s(\text{Green}) \cdot u(\$100)$, which implies

$s(\text{Blue}) > s(\text{Red})$.

Therefore, we reach a contradiction in the theory.

Cumulative prospect theory (Wakker, 2010) uses a nonlinear weighting function to account for this violation of subjective expected utility theory as follows.

If X is preferred to Y, then

$PU(X) = w(\text{Red}) \cdot u(\$100) > PU(Y) = w(\text{Blue}) \cdot u(\$100)$, which implies

$w(\text{Red}) > w(\text{Blue})$.

But if Y' is preferred to X', then

$PU(Y') = w(\text{Blue or Green}) \cdot u(\$100) > PU(X') = w(\text{Red or Green}) \cdot u(\$100)$,

which implies

$w(\text{Blue or Green}) > w(\text{Red or Green})$.

There is no contradiction if the weighting function is nonadditive:

$w(E_1 \text{ or } E_2) \neq w(E_1) + w(E_2)$.

There are, however, many other models and explanations for this phenomenon. One new idea is that risk and reward are negatively related in the real world. Bigger rewards are less likely to occur. Pleskac and Hertwig (2014) conducted an ecological analysis of life's gambles, ranging from the domains of roulette and life insurance to scientific publications and artificial insemination. Across all domains, payoffs and probabilities were negatively related. As a result, Pleskac and Hertwig (2014) argue that when confronted with the Ellsberg gambles, people might rely on a risk-reward heuristic and infer that the probability of winning for action Y with the unknown mixture is low when choosing between X and Y, but then infer that the probability of winning for action X′ with the unknown mixture is low when choosing between X′ and Y′, thus producing the reversal of preference.

2.7 Multiattribute Expected Utility Model

Finally, we come to the categories that combine multiattribute models with expected utility models. For an example experimental test of this model, see von Winterfeldt (1980), who found violations of the independence assumptions of the model. We will not go into the details here; instead, we recommend one of the most comprehensive treatments of this class of models, by Keeney and Raiffa (1993). Decision analysts have developed many sophisticated and user-friendly computer programs for conducting decision analyses using the multiattribute expected utility model.

2.8 Heuristic Decision Models

A completely different approach to decision-making is provided by using simple heuristic rules for making decisions. An example of a heuristic rule is the lexicographic (LEX) rule: Select the most important dimension and pick the option that is best on that dimension; if some options are nearly equivalent on the most important dimension, then drop the options that are clearly worse on the first dimension, but then compare the remaining options on the second-most-important dimension. This process continues until all the dimensions are processed. However, not all the dimensions need to be processed to make a decision, and the process can stop even on the first attribute.

Payne, Bettman, and Johnson (1993) propose that decision-makers possess a repertoire of strategies that they may apply to a decision, including noncompensatory rules that do not require trade-offs among attributes, such as LEX, as well as compensatory rules that are based on attribute trade-offs, such as a weighted additive rule or the expected utility rule. A strategy is selected by considering a trade-off between the mental effort required to apply the strategy and the accuracy or performance of the strategy. Thus, in situations involving time pressure, individuals may fall back on relatively simple strategies that do not involve complex calculations, such as LEX. In contrast, in important situations where a high level of performance is required, decision-makers may apply more cognitively intensive strategies, such as the weighted additive or expected utility rules.

Process-tracing methods can be used in computer controlled experiments to evaluate strategies empirically by presenting a table of options and attributes with the entries missing, and then have the decision-maker search for information by using a mouse to click on a missing entry to reveal the information; alternatively, all the information can be displayed, and eye movements can be used to record the information search (Payne et al., 1993). For example, the LEX rule implies that alternatives are compared within an attribute; in contrast, the weighted additive rule implies that attributes are evaluated within an alternative. These differences in processing sequence should be reflected in the pattern of information search.

For example, Brandstätter, Gigerenzer, and Hertwig (2006) recently proposed a LEX model called the "priority heuristic," which assumes the following process for positively valued gambles: (1) compare the lowest outcomes for each prospect, and if this difference exceeds a cutoff, then choose the best outcome based on this comparison; otherwise (2) compare the probabilities associated with the lowest payoffs, and if this difference exceeds a cutoff, then choose the best probability based on this comparison; otherwise (3) compare the maximum possible payoffs for each prospect and choose the best option. This model has been shown to account for many of the paradoxes of risky decision-making. However, it has also received some criticism for its failure to empirically account for choices among gambles (Birnbaum, 2008a).

2.9 Summary

Utility models are usually derived from a rational or axiomatic foundation. The main axioms are transitivity and independence. Transitivity is required to assign a real number to each alternative, and independence is required to produce the additive form. If one assumes that people are rational (at least in the manner assumed by these theories), then these utility models should be useful for predicting choice behavior.

A great deal of empirical work has been conducted to test the axioms of utility theories. This research finds systematic violations of the axioms in human choice behavior. However, this does not necessarily mean that we should throw out these utility models. They are simple and powerful and predict choice reasonably, even though they fall short in certain circumstances. Alternative models based on psychological rather than rational principles have been proposed to replace the rational models, such as cumulative prospect theory (Tversky & Kahneman, 1992) or, even more recently, Birnbaum's transfer attention exchange (TAX) model (Birnbaum, 2008b). The advantage of these psychological models is that they are able to account for a wider range of choice behavior, but at the cost of increased model complexity. As we shall see later in this book, utility theories still provide the core ideas for many of the choice models.

3 Random Utility Models

Human behavior is variable and somewhat unpredictable. Different people will make different choices because their utility functions, subjective probabilities, or both differ. Even within the same person, when facing the same choice on different occasions, the person will change his mind. This variability within a person can arise from a number of sources. The person may simply make an error on one occasion, they may focus attention on different features or attributes on different occasions, they might change with experience across occasions, the person's mood and motivation might be different at different points in time, it may be strategically useful to be unpredictable as in a game situation, and so on.

On the one hand, in some circumstances, we can use information about individual differences or changes in the environment or changes from past experiences in order to predict some of the variability in a person's choices. However, on the other hand, even after all the known information is used, there is a substantial amount of variance in choices that remains unpredictable. This unpredictability is almost always the case, we need to formulate probabilistic models of choice.

The utility models presented in chapter 2 are difficult to apply to probabilistic behavior because of their deterministic formulation. Somehow, they need to be generalized to account for this variability. One of the most important, useful, and applicable ways to do this is to reformulate the deterministic utility models as random utility models. The use of the word *random* here does not mean that choices are *completely* random; instead the term refers to the idea that utilities are treated as random variables with a distribution that could be biased to favor one option or another.

In this chapter, we first define a very generic random utility model and discuss its properties. Then we present the two most popular and important kinds of random utility models, the extreme value random utility type of model, and the Thurstone type of model, with normally distributed utilities. We end the chapter with a discussion of the advantages and disadvantages of the competing theories.

3.1 Generic Random Utility Model

We begin this discussion by assuming that there is a complete set T of N choice options $T=\{A_1, A_2, \ldots, A_N\}$. In some examples, we will label the options as $\{A, B, C\}$. These options might be consumer products, business investments, medical treatments, or other items. The decision-maker is presented with some subset $S \subset T$ of these N options. When a choice is made, a preference state ω is sampled from a sample space Ω, which can be used to determine the N real-valued utilities for each of the N options $U(\omega)=[U_1(\omega), \ldots, U_N(\omega)]$. In other words, the person is assigned a state that can be used to determine utilities for any of the N possible options. Once a preference state ω is sampled, the choice becomes deterministic. Behavior is probabilistic only because we do not know the state ω used to evaluate the options. Therefore, we have to assign the probabilities of sampling these states. This sampling process produces N random variables U_i, $i=1, \ldots, N$ (one for each option in T), which determine an N-dimensional random utility distribution function denoted by

$$F_T=p[U_1 \leq u_1, \ldots, U_N \leq u_N].$$

When the decision-maker is presented with a choice set S, they choose the option that has the maximum randomly sampled value in the choice set S, $\max\{U_i$, for $i \in S\}$. The choice probabilities for set S are based on the marginal dimensional distribution F_S, which is obtained by integrating F_T over the values of the options from the complete set T that are not presented in S. The probability that option A_i is chosen from the presented set S equals the probability that the randomly sampled value for the random variable U_i is the maximum:

$$p(A_i \mid S)=p[U_i=\max\{U_j, U_j \in S\}].$$

(It is assumed here that the probability of a tie is zero.) The main consistency assumption of random utility models is that the same N-dimensional distribution function F_T is used for all the choice sets. For example, if there are only three options $\{A, B, C\}$ in the choice set, the probability to choose option A equals

$$p(A \mid \{A, B, C\}) = p[U_A - U_B > 0 \cap U_A - U_C > 0].$$

So far, these models seem extremely general. Can they be empirically tested at this level of generality?

3.2 Empirical Properties of Random Utility Models

Transitivity As noted in chapter 2, transitivity is one of the main axioms of all the utility theories: if option A is preferred to B and option B is preferred to C, then option A is preferred to C. It seems that this would be an easy property to test empirically. But in reality, this turns out to be a fairly difficult empirical question. The problem concerns the interpretation of the preference relation. Does this preference relation refer to a single choice? This is not a good idea because transitivity could be violated accidentally by choice errors or other changes in preference across choice trials.

One better way to define a probabilistic version of preference is the majority choice definition. Following Luce (2000), we define the preference relation A is preferred or indifferent to B iff $p(A \mid \{A, B\}) \geq .50$. This leads to a definition of weak *stochastic* transitivity:

$$p(A \mid \{A, B\}) \geq .50,\ p(B \mid \{B, C) \geq .50 \rightarrow p(A \mid \{A, C\}) \geq .50.$$

This definition was examined by Tversky (1969), which we discuss later in this chapter.

The generic random utility class of models does not need to satisfy weak stochastic transitivity. Some special cases discussed later in this chapter do satisfy weak stochastic transitivity, but in general, this does not need to occur. The decision-maker can be transitive whenever a state ω is sampled, but on each occasion, the preference order is different and averaging across these different preference orders can violate weak stochastic transitivity (Regenwetter, Dana, & Davis-Stober, 2011). For instance, suppose that at different times, a person makes pairwise choices, and at these times, the person experiences a different state of ω. Further, suppose that the person experiences the following transitive preference states:

A preferred to B preferred to C—call this preference state *ABC*.

B preferred to C preferred to A—call this preference state *BCA*.

C preferred to A preferred to B—call this preference state *CBA*.

For example, with a pairwise choice occasion where A and B are presented, if the person happens to be in preference state *ABC*, then A will be chosen over B; but on another occasion where A and B are presented and the person is in preference state BCA, then B will be chosen over A. If we assume that these three preference states occur equally often, then we observe that

$$p(A \mid \{A, B\}) = 2/3,\ p(B \mid \{B, C\}) = 2/3,\ \text{but } p(A \mid \{A, C\}) = 1/3,$$

violating weak stochastic transitivity. This is called the *Condorcet paradox*. Thus, generic random utility models do not need to satisfy weak stochastic transitivity.

All random utility models, however, must satisfy another transitivity property called the *triangular inequality for binary choices* (Becker, DeGroot, & Marschak, 1963):

$$p(A \mid \{A, B\}) + p(B \mid \{B, C\}) - p(A \mid \{A, C\}) \leq 1,$$

or equivalently,

$$p(C \mid \{A, C\}) \leq p(C \mid \{B, C\}) + p(B \mid \{A, B\}).$$

This is the case for the following reason. Define $p(ABC)$ as the probability of a state in which there is a strict transitive order, $A > B > C$. The pairwise choices are determined from the probabilities of the preference states as follows:

$$p(C \mid \{B, C\}) = p(CBA) + p(CAB) + p(ACB)$$

$$p(B \mid \{A, B\}) = p(BAC) + p(BCA) + p(CBA)$$

$$p(C \mid \{A, C\}) = p(CAB) + p(CBA) + p(BCA),$$

which implies that $p(C \mid \{B, C\}) + p(B \mid \{A, B\}) =$

$$[p(CBA) + p(CAB) + p(ACB)] + [p(BAC) + p(BCA) + p(CBA)]$$
$$\geq [p(CAB) + p(CBA) + p(BCA)] = p(C \mid \{A, C\}).$$

Given the importance of the transitivity property, one would expect a firm resolution with regard to its empirical status. Tversky (1969) initially conducted a test of weak stochastic transitivity using a design illustrated in table 3.1.

The design in table 3.1 is a simplified version of Tversky's actual experiment, which involved six rather than three gambles. The intuition was that people generally think that probability of winning is more important than the amount to win. However, when comparing options A and B, the difference in probability is too small to worry about, but the difference in amount is noticeable (i.e., 50 percent greater), and so a person may choose option B

Table 3.1
Simplified version of Tversky's design.

Gamble	Probability of winning	Amount to win
A	.60	100$
B	.55	150$
C	.50	200$

more frequently than A. For a similar reason, the person may choose option C more frequently than B. However, when comparing options A and C, the difference in probability is now noticeable, and the person pays more attention to probability and chooses option A more frequently than C. The latter result produces a violation of weak stochastic transitivity.

We need to be careful here. The study was designed to test transitivity at the individual level (after all, different individuals can have different preference orders). In the actual experiment, each person was presented each pair of gambles in twenty trials (interleaved with other choices to encourage independent responses). This sample size produces sample proportions rather than population probabilities. Therefore, a statistical method must be developed to test whether the sample proportions statistically deviate from weak stochastic transitivity. Tversky (1969) developed a statistical test that revealed statistically significant violations from a majority of a specially selected group of participants (selected by a pretest to be vulnerable to violations of transitivity).

Later, however, the assumptions underlying the statistical test that Tversky derived were called into question, and a more appropriate statistical test was derived (Iverson & Falmagne, 1985). Based on this revised analysis, the conclusion was reversed and only a very small minority of participants violated weak stochastic transitivity.

More recently, Regenwetter et al. (2011) criticized the use of weak stochastic transitivity to test the transitivity axiom of random utility models because some random utility models are not required to satisfy weak stochastic transitivity. Regenwetter et al. (2011) developed nonparametric statistical tests of the triangular inequality and found that most participants did not produce statistically significant violations. In sum, recent evidence using adequate statistic tests shows that these violations appear to be the exception rather than the rule.

Regularity According to regularity, the addition of an option to a choice set should never increase the probability of selecting an option from the original set. For example, $p(A \mid \{A, B\})$ must be greater than or equal to $p(A \mid \{A, B, C\})$; adding option C should only compete and decrease the probability of choosing A.

All random utility models must satisfy regularity because it follows from the fact that

$$\begin{aligned} &p(U_A - U_B > 0 \cap (U_A - U_C > 0) \\ &\quad = p(U_A - U_B > 0) \cdot p(U_A - U_C > 0 \mid (U_A - U_B) > 0) \\ &\quad \leq p(U_A - U_B > 0). \end{aligned}$$

Violations of regularity were first found by Huber, Payne, and Puto (1982), which are called *asymmetric dominance effects* or *attraction effects*. The essential idea here is to examine the preferences for options (say, A versus B) with or without the context of another *deficient* option D. Options A and B are designed to be qualitatively different or dissimilar—for example, option A could be high in quality and price, whereas option B is low in quality and price. Option D is designed to be very similar to option A; but the key difference needed to produce an attraction effect is to make D deficient or defective compared to option A (e.g., D is of slightly lower quality but almost the same price as A, so D is nearly dominated by A). In this situation, D is rarely chosen, but including D in the choice set that includes A and B increases the probability of choosing option A, so $p(A \mid \{A, B, D\}) > p(A \mid \{A, B\})$. These findings have been replicated in many studies (e.g., Wedell, 1991), and they call into question all random utility models (a random utility model that assumes a single joint distribution of random utilities across all options is used to make choices from any subset of choice options).

3.3 Thurstone and Probit Utility Models

Thurstone's (1927) choice model was one of the first probabilistic choice models to be proposed. Originally, the theory was applied to psychophysics, but later, Thurstone (1959) applied his model to a wide range of problems, including preferential choice. Subsequently, there have been numerous applications and developments (Bockenholt, 2006).

Essentially, the Thurstone model (also called the *probit model*) assumes that the N-dimensional random utility vector U is multivariate normally distributed with a mean vector μ and a variance-covariance matrix Σ. Each coordinate μ_i of the mean vector represents the mean or expected value of the utility for an option:

$$\mu_i = \mathrm{E}(U_i).$$

The diagonal entries of Σ, σ_{ii}, represent the variances of each utility,

$$\sigma_{ii} = \mathrm{V}[U_i] = \mathrm{E}[(U_i - \mu_i) \cdot (U_i - \mu_i)],$$

and the off-diagonal entries, σ_{ij}, represent the covariance between a pair of utilities:

$$\sigma_{ij} = \mathrm{Cov}[U_i, U_j] = \mathrm{E}[(U_i - \mu_i)) \cdot (U_j - \mu_j)].$$

The mean utility of an option can be computed from a deterministic utility model, as discussed in chapter 2. The variance of the utilities represents a kind of uncertainty that a person can have regarding an option (MacKay & Zinnes 1986). Increasing variance represents increasing uncertainty, and options can differ on the degree of uncertainty. The covariance is important for capturing context effects produced by similarity between pairs of options.

Figure 3.1 illustrates how the Thurstone model works for binary choices. In the binary choice case, the probability of choosing option 1 over option 2 is determined by the difference in random utilities, $U_1 - U_2$, for the two options. For the moment, focus on the normally distributed case representing the difference in utility when the variance is relatively large. If this difference exceeds zero, then option 1 is chosen; otherwise, option 2 is chosen. In figure 3.1, option 1 would be chosen for all values that exceed the vertical line at zero. The probability of choosing option 1 is the area above the vertical line. For the high variance case, the probability is .69. Figure 3.1 also shows how the variance of the differences can moderate the choice probability. The narrow distribution represents a choice with a small standard deviation, and the other represents a choice with a large standard deviation. The mean differences are the same in the two cases (with the mean equal to 1 in both cases), but the choice probability changes from .69 to .98 for the large ($\sigma = 2$) and small ($\sigma = 1/2$) standard deviation cases, respectively.

Case V The simplest version of the Thurstone model is what is known as case V. Consider a binary choice between any pair of options, such as *A, B*. Suppose that each option corresponds to a normally distributed random

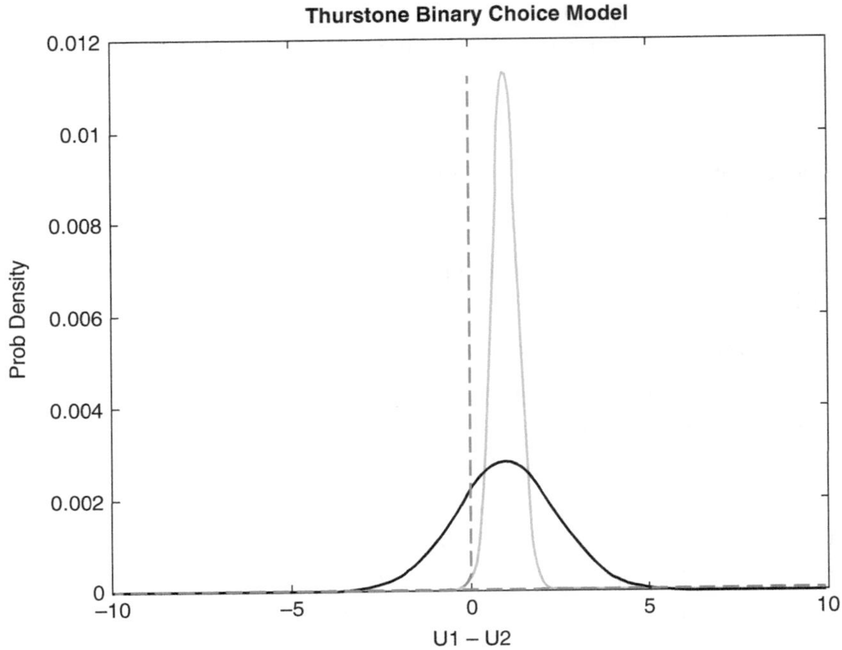

Figure 3.1
Illustration of Thurstone choice model. Horizontal axis represents the utility difference, vertical axis represents probability density. Area above the verical line determines probability to choose A. One distribution is shown with a choice pair with high variance, and a second is shown with a choice pair with low variance.

variable, such as U_A for option A and U_B for option B. Define the mean for each option as $\mu_i = \mathrm{E}[U_i]$ (e.g., option A has mean μ_A and option B has mean μ_B). Also assume that they all have the same common variance, σ^2. Finally, assume that all the random variables are uncorrelated. These assumptions reduce the general Thurstone model to a simple, case V Thurstone model. According to this model, for any pair,

$$p(A \mid \{A, B\}) = \Pr[U_A - U_B > 0] = F\left(\frac{\mu_A - \mu_B}{\sigma}\right), \text{ with } \sigma \text{ constant across pairs.}$$

Note that usually $\sigma = 1$, without loss of generality, because the parameters used to define the means can be adjusted to accommodate any single common value of σ. In sum, the choice probability is an increasing function of the difference between the means. The case V model is commonly used in applications, and it is available in statistical packages that include the generalized linear model with the probit link function.

Strong stochastic transitivity The case V model satisfies a probabilistic version of transitivity called *strong stochastic transitivity*. According to strong stochastic transitivity, if

$p(A \mid \{A, B\}) \geq .50$ and $p(B \mid \{B, C\}) \geq .50$, then

$p(A \mid \{A, C\}) \geq p(A \mid \{A, B\})$ and $p(A \mid \{A, C\}) \geq p(B \mid \{B, C\})$.

According to the case V model,

$p(A \mid \{A, B\}) \geq .50$ implies $\mu_A \geq \mu_B$

and

$p(B \mid \{B, C\}) \geq .50$ implies $\mu_B \geq \mu_C$,

and so we have $\mu_A \geq \mu_B \geq \mu_C$. The fact that μ_B is between μ_A and μ_C implies that

$p(A \mid \{A, C\}) \geq p(A \mid \{A, B\})$, because $\mu_A - \mu_C \geq \mu_A - \mu_B$

and

$p(A \mid \{A, C\}) \geq p(B \mid \{B, C\})$, because $\mu_A - \mu_C \geq \mu_B - \mu_C$.

Similarity effects in binary choice The assumption of strong stochastic transitivity is problematic because of similarity effects that occur with binary choice (Tversky & Russo, 1969; Rumelhart & Greeno, 1971). Consider the following example, taken from a study by Mellers and Biagini (1994), illustrated in table 3.2. When comparing laptop A to B, option A almost dominates B because the two options are very similar in terms of weight but A is much cheaper. Therefore, $p(A \mid \{A, B\})$ is close to 1.0. When comparing B to C, the comparison is very difficult because of the conflict produced by a large decrease in weight for C offset by a large increase in price for C. However, suppose that the change in price is considered slightly

Table 3.2
Design used by Mellers and Biagini.

Laptop	Weight	Cost
A	1.03 kg	\$1,000
B	1.02 kg	\$1,100
C	0.45 kg	\$2,000

more valuable so that $p(B \mid \{B, C\})$ is slightly greater than .50; and for the same reason (conflict between weight and price), suppose that $p(A \mid \{A, C\})$ is slightly greater than .50. This pattern produces a violation of strong stochastic transitivity because $p(A \mid \{A, B\}) > .50$ and $p(B \mid \{B, C\}) > .50$, but $p(A \mid \{A, C\}) < p(A \mid \{A, B\})$. The similarity between A, B makes this choice easy, and the conflict produced by A, C makes this choice difficult.

Binary choice independence The case V model also satisfies a property called *binary choice independence*. Binary choices satisfy a property called *independence* if the following property always holds for pairs of choices:

$$p(A \mid \{A, C\}) > p(B \mid \{B, C\}) \text{ implies } p(A \mid \{A, D\}) > p(B \mid \{B, D\}) .$$

The case V model satisfies this property because

if $p(A \mid \{A, C\}) > p(B \mid \{B, C\})$, then $\mu_A > \mu_B$, which then implies that

$p(A \mid \{A, D\}) > p(B \mid \{B, D\})$.

Variance effect Independence can be violated when the variance of the utilities changes across choice pairs. There is extensive empirical evidence for variance heterogeneity effects regarding choices between gambles (Blavatskyy, 2007; Hey & Orme, 1994; Erev & Barron, 2005; Busemeyer, 1985; Wilcox, 2011). Consider, for example, the following experiment by Busemeyer (1985). Option A was a gamble that produced normally distributed payoffs with a mean equal to zero and a standard deviation equal to 5 cents. Option B was a gamble producing normally distributed payoffs with a mean equal to zero and a standard deviation equal to 50 cents. Option C delivered a certain loss of 3 cents, and option D delivered a certain gain of 3 cents. The results found by Busemeyer (1985), accumulated across hundreds of choice trials for each person, were as follows:

$p(A \mid \{A, C\}) = .88 > p(B \mid \{B, C\}) = .62$, but

$p(A \mid \{A, D\}) = .11 < p(B \mid \{B, D\}) = .27$.

In other words, using C as a comparison, A seems to have a stronger preference over C than B over C; however, using D as a comparison, A seems to have a weaker preference over D than B over D. This violates binary choice independence. In this experiment, when choosing between a gamble and a sure thing, the choice probability was determined by the ratio of mean difference and standard deviation—but the standard deviation changed across the pairs according to the variance of the gamble. The variance was small for gamble A and large for gamble B, and this change in variance across choice pairs produced violations of the binary choice independence property.

A more general binary choice model To illustrate how the Thurstone model works more generally, consider again a choice between two options, A, B. Suppose that these are two consumer products described in terms of quality and price. Assume that the random utility for each option is determined by a weighted additive utility model as follows (cf. Hausman & Wise, 1978):

$$U_A = W_Q \cdot v_{AQ} + W_P \cdot v_{AP}$$

$$U_B = W_Q \cdot v_{BQ} + W_P \cdot v_{BP}.$$

Assume that the values of each alternative on each attribute, v_{ij}, are fixed equal to their objective values, but the two subjective weights are bivariate normally distributed:

$$W = \begin{bmatrix} W_Q \\ W_P \end{bmatrix} \sim N(w, \Psi),\ w = \begin{bmatrix} w_Q \\ w_P \end{bmatrix} = \begin{bmatrix} E[W_Q] \\ E[W_P] \end{bmatrix},$$

$$\Psi = \begin{bmatrix} \Psi_{QQ} & \Psi_{QP} \\ \Psi_{PQ} & \Psi_{PP} \end{bmatrix}$$

$$= \begin{bmatrix} E[(W_Q - w_Q)(W_Q - w_Q)] & E[(W_Q - w_Q)(W_P - w_P)] \\ E[(W_P - w_P)(W_Q - w_Q)] & E[(W_P - w_P)(W_P - w_P)] \end{bmatrix}.$$

The choice probability is determined by

$$p(A \mid \{A,B\}) = \Pr[U_A - U_B > 0] = F\left(\frac{\mu_A - \mu_B}{\sigma_{Diff}}\right),$$

where F is the cumulative normal distribution function, and

$$\mu_A = w_Q \cdot v_{AQ} + w_P \cdot v_{AP}$$

$$\mu_B = w_Q \cdot v_{BQ} + w_P \cdot v_{BP}$$

are the mean utilities for each option. The standard deviation is no longer constant across pairs, and instead it changes as follows [with $V(X)$ symbolizing the variance of the random variable X]:

$$\sigma_{Diff} = V[(U_A - U_B)] = V[W_Q \cdot (v_{AQ} - v_{BQ}) + W_P \cdot (v_{AP} - v_{BP})]$$

$$= V[W_Q \cdot (v_{AQ} - v_{BQ})] + V[W_P \cdot (v_{AP} - v_{BP})] + 2 \cdot \mathrm{Cov}[W_Q \cdot (v_{AQ} - v_{BQ}),\ W_P \cdot (v_{AP} - v_{BP})]$$

$$= (v_{AQ} - v_{BQ})^2 \cdot \psi_{QQ} + (v_{AP} - v_{BP})^2 \cdot \psi_{PP} + 2 \cdot (v_{AQ} - v_{BQ}) \cdot (v_{AP} - v_{BP}) \cdot \psi_{QP}.$$

Note that if the options are similar, such that $(v_{AQ} - v_{BQ})$ and $(v_{AP} - v_{BP})$ are close to zero, then the variance is reduced; but if the options are dissimilar, such that $(v_{AQ} - v_{BQ})$ and $(v_{AP} - v_{BP})$ are far from zero, then the variance is increased. Thus the variance changes across pairs in this model.

This binary choice model has five parameters: two contained in the mean weight vector w, and three contained in the symmetric variance covariance matrix Ψ. These five parameters can then be used to predict all of the various pairs of options generated from using different values assigned to the options. This model is also capable of accounting for the similarity effects reported by Mellers and Biagini (1994).

Triadic choice Now consider the Thurstone model for three choice options, $\{A, B, C\}$. In this case, we assume that the 3×1 random utility vector U is multivariate normally distributed with a 3×1 mean vector μ and a 3×3 variance covariance matrix Σ. Then the choice probabilities are determined as follows:

$$p(A \mid \{A, B, C\}) = p[U_A - U_B > 0\ ,\ U_A - U_C > 0],$$

$$p(B \mid \{A, B, C\}) = p[U_B - U_A > 0\ ,\ U_B - U_C > 0],$$

$$p(C \mid \{A, B, C\}) = p[U_C - U_A > 0\ ,\ U_C - U_B > 0].$$

Note that this requires integrating a bivariate normal distribution across the upper-right quadrant when both differences are positive. Numerical integration methods are needed to carry out this computation.

Choice independence The binary choice independence property (which is often related to the independence from irrelevant alternatives axiom) can be generalized for more than two choices as follows. Suppose that S and T are both sets of choice options containing the options A, B. A choice model satisfies choice independence if $p(A \mid S) \geq p(B \mid S)$ implies $p(A \mid T) \geq p(B \mid T)$. The case V model must satisfy this property because $p(A \mid S) \geq p(B \mid S)$ implies $\mu_A \geq \mu_B$, and holding the other means constant, $\mu_A \geq \mu_B$ implies that $p(A \mid T) \geq p(B \mid T)$. However, the general Thurstone model does not have to satisfy this property

Similarity effects with multiple choice In fact, Tversky (1972) empirically reports systematic violations of independence, with findings of similarity effects with trinary choices. Tversky used the following example to illustrate how this happens. Suppose that a person is given a choice between a CD by Beethoven (option B) and one by Miles Davis (option D), and the person likes both very much, but perhaps this person likes

Beethoven a little bit better, so that $p(B \mid \{B, D\}) = .60 > p(D \mid \{B, D\})$. Now suppose that the person is given a trinary choice among Beethoven, Miles Davis, and Mozart. The person might become torn between Mozart and Beethoven, but the chances of taking the Miles Davis CD remain the same, so $p(D \mid \{B, D, M\}) > p(B \mid \{B, D, M\})$. This is just a thought example, but Tversky (1972) empirically demonstrated similarity effects using choices among job candidates and choices among gambles.

The general Thurstone model does not have to satisfy choice independence. Suppose that three options $A1, A2, A3$ produce three random utilities that are normally distributed with

$$\mu = \begin{bmatrix} \mu_1 \\ \mu_2 \\ \mu_3 \end{bmatrix} = \begin{bmatrix} .42 \\ .56 \\ .58 \end{bmatrix}, \text{and } \Sigma = \begin{bmatrix} 10.2 & 2.6 & 1.8 \\ 2.6 & 8.8 & 7.4 \\ 1.8 & 7.4 & 10.2 \end{bmatrix}$$

Figure 3.2 illustrates the bivariate joint distributions produced for each choice option for the triadic choices. The left panel represents the joint distribution for the choice of option $A1$ over $A2$ and $A3$. The middle panel is for option $A2$, and the right panel is for option $A3$. In the left panel, the horizontal axis represents the difference in utility between options $A1$ and $A3$, and the vertical axis represents the difference between options $A1$ and $A2$. Within each panel, there is a two-dimensional density distribution, which would be shaped like a hill coming off the page. We do not show the density distribution; instead, we show the equal density contours that it generates. Each contour is an elliptical curve the represents a slice into the hill at some fixed height of the hill. The blue indicates lower density (lower height) and the red indicates higher density. The angle of the ellipse represents the correlation between the pairs. The probability of choosing *the designated option in each panel* is the area in the upper-right quadrant above zero on each coordinate. The positive correlation rotates the ellipse into the positive quadrant. In the middle panel, the correlation is near zero. In the middle panel, the wider range on the ellipse in the vertical direction indicates a larger variance for this difference between utilities for options $A2$ and $A1$; the narrow range in the horizontal direction indicates a smaller variance in the utility difference between options $A2$ and $A3$. The distributions shown in figure 3.2 produce the following choice probabilities:

$$p(A1 \mid \{A1, A2\}) = .49,\ p(A1 \mid \{A1, A3\}) = .48,\ p(A3 \mid \{A2, A3\}) = .50$$

$$p(A1 \mid \{A1, A2, A3\}) = .40,\ p(A2 \mid \{A1, A2, A3\}) = .27,\ p(A3 \mid \{A1, A2, A3\}) = .33.$$

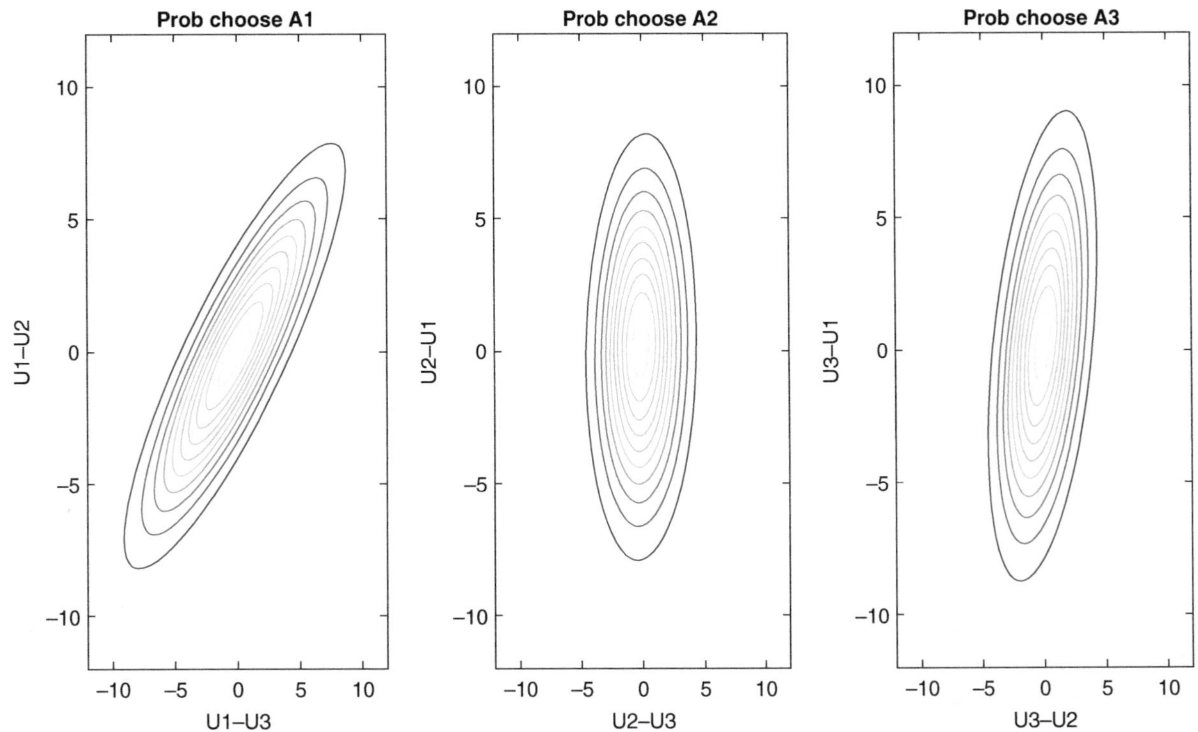

Figure 3.2
Bivariate joint distributions produced for each option for the triadic choice among options A1, A2, and A3. See color plate 1.

As can be seen from these calculations, this more general Thurstone model, with a full variance-covariance matrix, violates choice independence. Option $A1$ is chosen less frequently than option $A3$ in binary choice, but option $A1$ is chosen most frequently in the triadic choice. This is because options $A2$ and $A3$ are more highly correlated with each other compared to option $A1$.

General properties Like all random utility models, the general Thurstone model has to satisfy regularity. This model does not have to satisfy strong stochastic transitivity, but it does have to satisfy moderate stochastic transitivity (Halff, 1976), which states that if $p(A \mid \{A, B\}) \geq .50$ and $p(B \mid \{B, C\}) \geq .50$, then $p(A \mid \{A, C\}) \geq \min \{p(A \mid \{A, B\}), p(B \mid \{B, C\})\}$. This implies that the Thurstone model must also satisfy weak stochastic transitivity. Finally, the general Thurstone model does not have to satisfy the choice independence property.

3.4 Logistic, Extreme Value, and Luce Choice Models

Logistic choice model A very popular choice model is the *logistic choice model* (Louviere, Hensher, & Swait 2000). Suppose that two consumer products are described by a common set of attributes (e.g., quality and cost). The value of a consumer product i on an attribute j is denoted as v_{ij}. Using a weighted additive model, the value of the consumer product is the weighted sum of the attribute values $u_i = \Sigma_j \, w_i \cdot v_{ij}$. Now suppose that consumers are given a binary choice between consumer products A, B. According to the logistic choice model, the probability to choose A over B equals

$$p(A \mid \{A, B\}) = \frac{1}{1 + e^{-\gamma \cdot (u_A - u_B)}},$$

where $\gamma > 0$ is a parameter that determines the steepness of the S-shaped function that maps the utility difference into choice probability. We can transform the ratio of choice probabilities into a logit score as follows:

$$\frac{p(A \mid \{A, B\})}{p(B \mid \{A, B\})} = e^{\gamma \cdot (u_A - u_B)}$$

$$\ln \frac{p(A \mid \{A, B\})}{p(B \mid \{A, B\})} = \gamma \cdot (u_A - u_B) = \gamma \cdot \sum_j w_j \, (v_{Aj} - v_{Bj}).$$

This model can be rewritten as

$$\frac{e^{\gamma\cdot(u_A - u_B)}}{e^{\gamma\cdot(u_A - u_B)}}\cdot\frac{1}{1+e^{-\gamma\cdot(u_A - u_B)}}=\frac{e^{\gamma\cdot(u_A - u_B)}}{1+e^{\gamma\cdot(u_A - u_B)}}$$

$$=\frac{e^{\gamma\cdot u_A}e^{-\gamma\cdot u_B}}{1+e^{\gamma\cdot u_A}e^{-\gamma\cdot u_B}}=\frac{e^{\gamma\cdot u_A}}{e^{\gamma\cdot u_B}+e^{\gamma\cdot u_A}}=\frac{v(A)}{v(B)+v(A)},$$

The last version is called the *ratio of strengths model*. Therefore, when working with binary choices, the ratio of strengths model can be transformed into a logistic choice model. The logistic choice model is a very popular model in the choice modeling literature, and it is available in many commonly used statistical packages that include the generalized linear model with the logit link function. This popular logistic is a special case of the more general Luce choice model, which in turn is a special case of the more general extreme value utility model (McFadden, 1981).

Luce choice model Luce (1959) introduced a choice model frequently called the ratio of strengths model. This model is popular among researchers in many fields, including neural network models and reinforcement learning models (where it is often called the *soft max rule*). According to this model, a strength $v(A_i)$ is assigned to each option, and

$$p(A_i \mid S)=\frac{v(A_i)}{\sum_{j\in S} v(A_j)}.$$

A strong assumption of this theory is that the value assigned to an option depends only on that option, and not on the other options presented in the choice set. In other words, the value of an option is context free.

The ratio of strengths model satisfies regularity because adding new options to the choice set increases the denominator (for $S \subset T$) as follows:

$$p(A_i \mid S)=\frac{v(A_i)}{\sum_{j\in S} v(A_j)}\geq\frac{v(A_i)}{\sum_{j\in T} v(A_j)}=p(A_i \mid T).$$

Extreme value random utility models So far, the Luce (1959) choice model does not state anything about random utilities. In fact, the ratio of strengths model was originally described as a probabilistic process of fixed strength. However, it was proved later that the ratio of strengths model could be derived from an extreme value random utility distribution (Yellot, 1977; McFadden, 1981). For a single option A_i with strength $v(A_i)$, the extreme value cumulative distribution is defined as

$$p[U_i < u] = F(u) = exp[v(A_i) \cdot u].$$

Note that $u \le 0$ is needed to guarantee that the cumulative distribution equals 1 at the maximum possible value. The density is obtained from the derivative

$$dF(u)/du = f(u) = v(A_i) \cdot exp[v(A_i) \cdot u].$$

If we assume statistically independent extreme value random utilities, then

$$\begin{aligned} p[U_i &= \max\{U_j, j \in S\}] \\ &= \int_{-\infty}^{0} f_i(u) \cdot \prod_{j \neq i} \Pr[U_j < u] du \\ &= \int_{-\infty}^{0} v_i e^{v_i \cdot u} \cdot \prod_{j \neq i} e^{v_j \cdot u} du \\ &= v_i \int_{-\infty}^{0} \prod_{j} e^{v_j \cdot u} du = v_i \int_{-\infty}^{0} \exp\left(u \cdot \sum_j v_j \right) du \\ &= \frac{v_i}{\sum_j v_j} \cdot \exp\left(u \cdot \sum_j v_j \right) \Bigg|_{u=-\infty}^{u=0} = \frac{v_i}{\sum_j v_j}. \end{aligned}$$

McFadden's (1981) generalized extreme value model is a generalization of the ratio of strengths model, in which the extreme value random utilities are not assumed to be independent. This is important because the correlations between random variables produce a model that can capture the effects of context on choice. Later in this chapter, we present a model called the elimination by aspects model (Tversky, 1972a, b), which is very similar to McFadden's model. For now, though, we continue to examine the properties of the ratio of strengths model.

Choice axiom The Luce choice model implies the following choice axiom. It describes the probability of choosing option A_i from a larger set T in terms of the product of the probability of choosing an option from a subset S of the larger set, and the probability of choosing A_i from the subset S:

$$p(A_i \mid T) = p(S \mid T) \cdot p(A_i \mid S),$$

where $p(S \mid T)$ is the probability of choosing one of the options in the subset S when presented with the larger choice set T. This follows from the ratio of strengths rule:

$$p(A_i \mid T) = \frac{v(A_i)}{\sum_{j \in T} v(A_j)} = \frac{\sum_{j \in S} v(A_j)}{\sum_{j \in T} v(A_j)} \cdot \frac{v(A_i)}{\sum_{j \in S} v(A_j)}.$$

What is surprising is that Luce (1959) also proved the implication, but in the other direction. If a person obeys the choice axiom, then we can always construct a ratio of strengths model to reproduce their choice probabilities. One implies the other.

Binary choices also obey the following product rule:

$$\frac{p(A\mid\{A,B\})}{p(B\mid\{A,B\})}\cdot\frac{p(B\mid\{B,C\})}{p(C\mid\{B,C\})}=\frac{v(A)}{v(B)}\cdot\frac{v(B)}{v(C)}=\frac{v(A)}{v(C)}=\frac{p(A\mid\{A,C\})}{p(C\mid\{A,C\})}.$$

It turns out that the product rule does not describe choice behavior very well because it fails to account for similarity effects that vary across choice sets. The following thought experiment was initially proposed in a comment to Luce's book by DeBreu. Consider a choice between Rome and Paris. Suppose that you like both, so the probability of choosing Paris is approximately .50. Now consider a choice between the exact same trip to Paris plus a dollar coupon and Rome. The dollar coupon is too small to have any effect on your preference, so the probability remains .50. These hypothetical results imply

$$\frac{v(Paris+\$)}{v(Rome)}\cdot\frac{v(Rome)}{v(Paris)}=\frac{v(Paris+\$)}{v(Paris)}\approx 1.$$

The product rule implies that the probability of choosing Paris plus a dollar coupon over the same trip to Paris, but without the coupon, is approximately .50 (maybe a tiny bit more); however, the empirical probability is expected to be near 1.0 because Paris plus a dollar dominates Paris without the dollar.

Strong stochastic transitivity The ratio of strengths model satisfies strong stochastic transitivity. According to the ratio of strengths model,

$$p(A\mid\{A,B\})\geq .50 \text{ implies } v(A)\geq v(B)$$

and

$$p(B\mid\{B,C\})\geq .50 \text{ implies } v(B)\geq v(C).$$

Thus we have $v(A)\geq v(B)\geq v(C)$, such that

$$p(A\mid\{A,C\})\geq p(A\mid\{A,B\}),$$

because $v(B)\geq v(C)$ increases the denominator of $p(A\mid\{A,B\})$ and

$$p(A\mid\{A,C\})\geq p(B\mid\{B,C\})$$

because $v(A)\geq v(B)$ increases the numerator for $p(A\mid\{A,B\})$. As previously mentioned with the case V Thurstone model, the assumption of strong stochastic transitivity is problematic because of similarity effects (Tversky & Russo, 1969; Rumelhart & Greeno, 1971). Similarity effects imply that the value assigned to an option depends on what other options are presented in the choice set. In other words, values are context dependent.

Choice independence axiom The choice axiom immediately implies another fundamental property discussed earlier in this chapter—namely, the choice independence axiom (also sometimes related to the independence from irrelevant alternatives axiom). Note that if

$$p(A \mid \{A, B\}) > p(B \mid \{A, B\}), \text{ then } v(A) > v(B),$$

which then implies

$$\frac{v(A)}{v(A)+v(B)+v(C)} > \frac{v(B)}{v(A)+v(B)+v(C)},$$

which finally implies

$$p(A \mid \{A, B, C\}) > p(B \mid \{A, B, C\})\ .$$

In other words, the preference order for A, B, as measured by choice probability, remains the same across larger choice sets containing A, B.

Reference point effects As mentioned earlier, the choice independence assumption is problematic because choices tend to be contextual and violate this independence property. Similarity effects, as discussed previously, provide one type of finding. Reference point effects produce another type (Tversky & Kahneman, 1991). Consider the following example. Suppose that a person has commodity R_A but is given a choice to keep R_A or trade it for A or B. Suppose that A has a small advantage over R_A on the first dimension and a little disadvantage on the second; and suppose that B has a large advantage over R_A on the first dimension but a large disadvantage on the second. In this case, it is found that

$$p(A \mid \{A, B, R_A\}) > p(B \mid \{A, B, R_A\}).$$

Now suppose that a person has commodity R_B but is given a choice to keep R_B or trade it for A or B. However, suppose that in this case, the relations are reversed: B has a small advantage over R_B on the first dimension and a little disadvantage on the second; suppose that A has a large advantage over R_B on the first dimension but a large disadvantage on the second. In this case, it is found that

$$p(A \mid \{A, B, R_B\}) < p(B \mid \{A, B, R_B\}).$$

The reversal produced by the location of the reference constitutes a violation of choice independence.

3.5 Elimination by Aspects Model and Restle's Choice Model

Tversky (1972) also developed a random utility model that assumed correlation among the random utilities. The elimination by aspects model is also a generalization of the Luce choice model. Actually, the elimination by aspects model was not originally formulated as a random utility model, but it was later proved to be one, with correlated utilities.

Tversky (1972) initially built the elimination by aspects model from earlier ideas developed by Restle (1961). Restle wished to avoid the idea that options are necessarily represented by numerical dimensions. Instead, he represented options as a general set of aspects. Using this idea, Restle originally developed a binary choice model that was an important modification of Luce's model. When presented with a choice between options A, B, the aspects can be divided into three sets: aspects common to both options, aspects unique to A, and aspects unique to B. We denote the value of the unique aspects of A as $v(A \cap {\sim}B)$, and likewise, the value of the unique aspects of B is denoted $v({\sim}A \cap B)$. The common aspects have no effect on choice. The choice probability is determined by

$$p(A \mid \{A,B\}) = \frac{v(A \cap \sim B)}{v(A \cap \sim B) + v(\sim A \cap B)}.$$

Note that this model makes the values context dependent. In this way, it accounts for the similarity effect. For example, it solves the "Paris plus a dollar" problem because if we set A equal to Paris plus a dollar, and B equal to the same trip to Paris but without the dollar, then the common aspects are represented by Paris, $v(A \cap {\sim}B)$ is positive (representing the extra dollar) and $v({\sim}A \cap B)$ is zero (there is nothing unique about B because both options offer the same trip to Paris). Thus, the Restle model is sensitive to similarity between options. The Restle model is contextual because the value assigned to an option now depends on the other option (after eliminating common features).

Tversky (1972) extended the Restle model from binary choice to multialternative choice by proposing the following heuristic choice process: Pick an aspect, and then eliminate all options that do not contain the aspect;

continue this process until only one option remains. Let us consider a choice among three options $\{A, B, C\}$. The valued aspects include

$$v(A \cap \sim B \cap \sim C) = v_A,\ v(\sim A \cap B \cap \sim C) = v_B,\ v(\sim A \cap \sim B \cap C) = v_C,$$

$$v(A \cap B \cap \sim C) = v_{AB},\ v(A \cap \sim B \cap C) = v_{AC},\ v(\sim A \cap B \cap C) = v_{BC},$$

and $v(A \cap B \cap C)$, but the latter has no effect.

The choice probabilities for the subsets and the entire set are

$$p(A \mid \{A, B\}) = \frac{v_A + v_{AC}}{v_A + v_{AC} + v_B + v_{BC}}$$

$$p(A \mid \{A, C) = \frac{v_A + v_{AB}}{v_A + v_{AB} + v_C + v_{BC}}$$

$$p(B \mid \{B, C\}) = \frac{v_B + v_{AB}}{v_B + v_{AB} + v_C + v_{AC}}$$

$$p(A \mid \{A, B, C\}) = \frac{v_A + v_{AB} \cdot p(A \mid \{A, B\}) + v_{AC} \cdot p(A \mid \{A, C\})}{v_A + v_B + v_C + v_{AB} + v_{AC} + v_{BC}}$$

$$p(B \mid \{A, B, C\}) = \frac{v_B + v_{AB} \cdot p(B \mid \{A, B\}) + v_{BC} \cdot p(B \mid \{B, C\})}{v_A + v_B + v_C + v_{AB} + v_{AC} + v_{BC}}$$

$$p(C \mid \{A, B, C\}) = \frac{v_C + v_{AC} \cdot p(C \mid \{A, C\}) + v_{CB} \cdot p(C \mid \{C, B\})}{v_A + v_B + v_C + v_{AB} + v_{AC} + v_{BC}}.$$

To see how this model produces similarity effects, suppose that we have a choice set including a Lexus (luxury car), an Infinity (luxury car), and a Ford Focus (economy car). Suppose that the following values are assigned to the aspects:

$v_L = 1$ (unique style), $v_I = 2$ (unique gadget), $v_F = 12$ (economy), and $v_{LI} = 13$ (quality); and assume that the rest are zero. Then the choice probabilities are found to be

$$p(L \mid \{L, F\}) = 14/26 = .54,\ p(L \mid \{L, I\}) = 1/3 = .333,\ p(I \mid \{I, F\}) = 15/27 = .56$$

$$p(L \mid \{L, I, F\}) = [1 + 13 \cdot (1/3)]/28 = .19,\ p(I \mid \{L, I, F\}) = [2 + 13 \cdot (2/3)]/28 = .38$$

$$p(F \mid \{L, I, F\}) = 12/28 = .43.$$

From this example, we can see that although both Lexus and Infinity are chosen more often over the Ford in a binary choice, the Ford is chosen most frequently in the triadic choice, producing a violation of independence. The similarity between the Lexus and Infinity produces a competition between these two, which reduces their individual proportions in the triadic choice set, whereas the dissimilar Ford changes little in probability from binary to the triadic choice sets. This is how the elimination by aspects model accounts for similarity effects.

In general, the elimination by aspects model does not have to satisfy either independence or strong stochastic transitivity. By allowing correlated random utilities, which translate into various patterns of common and distinctive features, the elimination by aspects model can account for similarity effects on choice. However, it is incapable of accounting for violations of regularity produced by the attraction effects or for all types of violations of independence. In particular, Tversky and Kahneman (1991) had to invoke the concept of loss aversion to account for the reference point effect. If all the random utilities are assumed to be independent, which translates into all distinctive features and no common features, then the elimination by aspects model reduces to the Luce choice model. The advantage of the elimination by aspects over the Luce choice model is its ability to account for similarity effects; but the disadvantage is the increase in number of parameters (represented by the valued sets of features).

3.6 Summary and Conclusions

Table 3.3 summarizes the models and behavioral choice properties reviewed in this chapter. Each row represents a model discussed in this chapter, each

Table 3.3
Comparison of Choice models.

	WST	SST	Triangle inequality	Regularity	Independence
RUM			*x*	*x*	
Thurstone case V	*x*	*x*	*x*	*x*	*x*
Thurstone general	*x*		*x*	*x*	
Luce choice	*x*	*x*	*x*	*x*	*x*
Logistic choice	*x*	*x*	*x*	*x*	*x*
EBA	*x*		*x*	*x*	

RUM: generic random utility model
SST: strong stochastic transitivity
WST: weak stochastic transitivity
EBA: elimination by aspects

column represents a choice property, and x indicates that the model must satisfy the property.

The Thurstone case V model and the Luce choice model are simple models with few parameters. The Luce model is especially easy to apply to multiple (i.e., more than two) options. However, these simple models must satisfy some highly restrictive properties, including strong stochastic transitivity and choice independence. These properties are known to fail in many choice experiments, especially situations that vary similarity/correlation among options or the uncertainty/variance of the utility of options. Thus, the usefulness of these simple theories is limited to situations for which similarity is not a factor or variability does not change across options. More general models, such as the general Thurstone model, the McFadden model with correlated random utilities, or Tversky's elimination by aspect model, can account for violations of strong stochastic transitivity and choice independence, but at the cost of additional parameters. All the models in this chapter satisfy regularity, which is violated by the attraction effect. Ultimately, the choice of model depends on the purpose of the application and the extent to which a property of a model is violated in the application.

4 Statistical Methods

Choice models always contain parameters that need to be estimated from the data. For example, the weights of a weighted additive model or the risk aversion parameter of an expected utility model usually need to be estimated from the data. Furthermore, we need to compare the predictive accuracy of competing models, such as comparing the predictions of various types of random utility models. The purpose of this chapter is to review some of the statistical tools needed to estimate model parameters and compare choice models. More details can be found elsewhere (e.g., Busemeyer & Diederich, 2010; Lewandowsky & Farrell, 2011; Busemeyer, Wang, Townsend, & Eidels, 2015).

4.1 The Disjunction Effect

Here, we will analyze choice data collected in a study by Busemeyer, Matthew, and Wang (2006). They replicated and extended some earlier research by Shafir and Tversky (1992) that examined an interesting finding, called the *disjunction effect.*

Consider the prisoner's dilemma (PD), in which there are two players opposing each other, and each has two actions: cooperate or compete. An example payoff matrix, conditioned on each pair of actions, is shown in table 4.1. This is a one-shot game—two players play each other only *once.* Given each action by the opponent, the player is always better off to defect, and the same is true for the opponent; therefore, defect–defect is the dominant strategy.

In the standard version of the game, hereafter referred to as the *unknown condition,* the players simultaneously select an action without knowledge

Table 4.1
Example of the PD game.

	Player defects	Player cooperates
Opponent defects	You: 10 Other: 10	You: 5 Other: 25
Opponent cooperates	You: 25 Other: 5	You: 20 Other: 20

of the opponent's action. Two new manipulations were used to examine the disjunction effect: In one case, a player was initially informed that the opponent has chosen to compete; and in another case, the player was initially informed that the opponent has chosen to cooperate. This manipulation is designed to test the "sure thing" principle that lies at the foundation of utility theory (Savage, 1972/1954): if a person prefers to defect knowing that the opponent will defect *and* knowing that the opponent will cooperate, then the person should prefer to defect even when the opponent's choice is unknown.

Shafir and Tversky (1992) found that many players chose to defect knowing that the other player will defect, and they also chose to defect knowing that the other player chose to cooperate; however, they changed their minds and cooperated when they did not know the choice of the other player. Shafir and Tversky (1992) originally obtained this result by misinforming their participants that the opponent was actually a human player in another room communicating via computer. Significantly, Busemeyer, Matthew, and Wang (2006) replicated their findings even when participants were truthfully informed that the opponent's moves were chosen by a computer.

Busemeyer, Matthew, and Wang (2005) recruited eighty-eight graduate and undergraduate students at Indiana University who participated for monetary payoffs. Players could earn a minimum of $5, and they could earn up to $20 depending on the outcomes of their decisions.

Each participant played forty games on a computer. Eighteen of these were PD games, similar to that shown in table 4.2, and the other games were fillers (not analyzed). Six PD matrices were used. Each of the PD matrices was presented on three trials: one with the opponent's action unknown, one with the opponent's action known to be to defect, and one with the opponent's action known to be to cooperate (randomly intermixed with other games).

The six PD matrices described in table 4.2 were used. Each row of table 4.2 represents a different payoff condition (except the last row, described later

Table 4.2
Payoff matrices and mean choice proportions.

Matrix	D-D	C-D	D-C	C-C	P(D)	P(D | D)	P(D | C)
1	10, 10	5, 25	25, 5	20, 20	.66	.93	.82
2	5, 5	0, 20	20, 0	15, 15	.67	.91	.83
3	15, 15	10, 30	30, 10	25, 25	.64	.92	.84
4	8, 8	4, 20	20, 4	16, 16	.59	.89	.86
5	5, 5	0, 15	15, 0	10, 10	.78	.94	.85
6	2, 2	1, 6	6, 1	4, 4	.61	.90	.84
Mean					.66	.92	.84

in this chapter). The first four columns indicate the payoffs for the (player, computer) combination of actions, where the first payoff is for the player and the second is for the computer. The columns show (in order) the payoffs when the player and the opponent both defect (D-D), the player cooperates and the opponent defects (C-D), the player defects and the opponent cooperates (D-C), and finally when the player and the opponent both cooperate (C-C). The next three columns show the proportion of defect choices pooled across the eighty-eight participants for three conditions: the player did not know the opponent's choice [the column headed "P(D)"], the player was informed that the opponent defected [the column headed "P(D | D)"], and the player was informed that the opponent cooperated [the column headed "P(D | C)"]. The last row shows the averages across the six payoff conditions for the three scenarios.

The disjunction effect can be seen by comparing the P(D) column to the P(D | D) and P(D | C) columns. Note that both P(D | D) and P(D | C) far exceed P(D). In other words, most participants defected when the opponent's action was known (regardless of whether the opponent cooperated or defected), but many of these switched to cooperation when the opponent's action was unknown. Next, we build some simple choice models for these results.

4.2 Choice Models for the PD Game

Here, we formulate some choice models for this game based on ideas related to epistemic game theory (e.g., Brandenburger, 2014). The basic idea is that the player predicts the opponent's action and chooses the best response.

Model A: Suppose that a player believes with probability $0 \leq w \leq 1$ that the opponent will defect, and $(1-w)$ is the probability that the opponent will cooperate. When the player is informed that the opponent defected, then $w=1$; when the player is informed that the opponent cooperated, then $w=0$; and for the unknown case, w is a free parameter that we need to estimate by fitting the model to the data.

Define u_{ij} as the utility of the payoff that the player receives if the player takes strategy $i \in \{D, C\}$ and the opponent takes strategy $j \in \{D, C\}$. We will use a power function, $u(x)=x^c$, to represent the utility u of a payoff x. For example, using matrix 1 in table 4.2, if $i=D$ and $j=C$, then the payoff is 25 and the utility is $u_{DC}=25^c$; if the parameter c is set to .5, then the utility equals $u_{DC}=25^c=25^{.5}=5$. The parameter c will be estimated by fitting the model to the data. If $c<1$, then the utility function is risk averse; if $c=1$, then the utility function is risk neutral; and if $c>1$, then the utility function is risk seeking.

The expected utility that the player receives for defecting, then, equals

$$EU(\text{player defects}) = w \cdot u_{DD} + (1-w) \cdot u_{DC},$$

and the expected utility for the player for cooperating equals

$$EU(\text{player cooperates}) = w \cdot u_{CD} + (1-w) \cdot u_{CC}.$$

The difference in expected utility equals

$$D_p = EU(\text{player defects}) - EU(\text{player cooperates}).$$

If the players were deterministic, then when $D_p>0$, they would always defect. However, the players that we observed were not deterministic, so we need to employ a probabilistic choice model. A simple and common assumption is a logistic choice model:

$$p(\text{player defects}) = 1/(1+\exp(-a \cdot D_p)),$$

with a being another free parameter that we need to fit to the data. We will call this Model A, and it has three free parameters (w, a, and c).

Model B: Next, we will consider a second model to account for the disjunction effect. This model is based on the same assumptions as Model A, except that we add a bias, b, for defecting when the opponent's choice is known, such that the total utility for defecting is $D_b = a \cdot D_p + b$. Then the probability that the player defects equals

$$p(\text{player defects}) = 1/(1+\exp(-D_b)).$$

We will call this Model B, and it has four parameters (w, a, b, and c). Model A is nested in Model B because Model A is a special case, with $b=0$. Model B must fit at least as well, or even better than, Model A.

Model C: Finally, we form another model based on fairness considerations (Fehr & Schmidt, 1999). Suppose that the player considers not only the payoff they receive, but also the payoff for their opponent. Define v_{ij} as the utility of the payoff that the opponent receives if the player takes strategy $i \in \{D, C\}$ and the opponent takes strategy $j \in \{D, C\}$. For example, using matrix 1 in table 4.2, if $i = D$ and $j = C$, then $v_{CD} = 5^c$. The expected utility for the opponent when the player defects, then, equals

$$\text{EU(player defects)} = w \cdot v_{DD} + (1 - w) \cdot v_{DC},$$

and the expected utility for the opponent when the player cooperates equals

$$\text{EU(player cooperates)} = w \cdot v_{CD} + (1 - w) \cdot v_{CC}.$$

The difference equals $D_o = \text{EU(player defects)} - \text{EU(player cooperates)}$. We assume that the player's choice depends on both their utility difference, D_p, and the utility difference for the opponent, D_o, by combining them into the linear combination $D_c = a \cdot D_p + d \cdot D_o$. The new parameter d is estimated by fitting the model to the data. For this model, the probability that the player defects equals

$$p(\text{player defects}) = 1/(1 + \exp(-D_c)).$$

We will call this Model C, and note that Model A is also nested in Model C, because it is a special case of Model C with $d = 0$. But Model B is not nested in Model C, and vice versa.

4.3 Estimating Parameters and Comparing Models

Each model has a vector containing several parameters, and in general, we will call this vector θ_i for person i. In particular, $\theta = (w, a, c)$ for Model A, $\theta = (w, a, b, c)$ for Model B, and $\theta = (w, a, c, d)$ for Model C for each person. We do not know θ_i, so we need to estimate this from the data. Suppose that we select some values for θ_i for a given model and person. Then we can compute the probability of defection using the previously discussed formulas for each of the eighteen trials for the person. We simply plug in the payoffs for the payoff matrix presented in a trial into the expected utility formula to compute the difference D for a model, and then we insert the difference D into the choice probability function in order to compute the probability of defection for a trial. Define $p_i(t) = p(\text{player } i \text{ defects on trial } t \mid \theta_i, M) = 1/(1 + \exp(-D))$ as the predicted probability of defection on trial

t for participant i, and $q_i(t) = 1 - p_i(t)$ as the predicted probability of cooperation, given that we are using model M and we have selected a specific parameter vector θ_i to compute D.

We need to evaluate how good or bad our prediction is by comparing the predicted probability on each trial for a person to the observed choice on that same trial. Define $o_i(t)$ as a binary-valued random variable that equals $o_i(t) = 1$ in a trial when we observe that person i chooses to defect, and zero otherwise. We will assume that this random variable is independent across trials conditioned on (1) the person, (2) the payoff matrix presented on that trial, and (3) information about the opponent's action. Independence might be reasonable here because the payoff conditions included other types of games (e.g., chicken game, battle of sexes game, etc.) between the PD game trials, and all the payoff conditions were randomized across trials, with the constraint that a PD game was always separated by other types of games and other PD games.

Across the eighteen trials for a given person, we observe a pattern of one's and zeros that represent the eighteen choices. Define the observed choice vector for person i as $O_i = [o_i(t),\ i = 1,2,\ldots,18]$. According to a model M and a selected parameter vector θ_i, the log likelihood of observing the data vector O_i for person i for that model equals (assuming independent responses)

$$LL_i = \ln p(O_i \mid M, \theta_i) = \sum o_i(t) \cdot \ln(p_i(t)) + (1 - o_i(t)) \cdot \ln(q_i(t)),$$

where the sum extends across the eighteen trials for person i. The log likelihood provides a goodness of fit measure of the model to the data for the person. The total log likelihood across all eighty-eight participants, then, equals

$$TLL = \sum \ln p(O_i \mid M, \theta_i).$$

It is common to transform TLL into a badness of fit statistic called $G^2 = -2 \cdot TLL$, because G^2 can be used to perform model comparisons.

One method for comparing nested models is a chi-square test. Consider some arbitrary models labeled Model1 and Model2. Suppose that Model1 has p_1 parameters and it is nested within Model2, which has $p_2 > p_1$ parameters (e.g., Model1 is a special case of Model2), so that G^2(Model1) must be larger than G^2(Model2). Then the difference has a chi-square distribution:

$$G^2(\text{Model1}) - G^2(\text{Model2}) \sim \chi^2(p_2 - p_1),$$

with $p_2 - p_1$ degrees of freedom. If the observed chi-square difference exceeds a critical value defined by the .95 percentile for the chi-square distribution, then we reject the null hypothesis that there is no difference between the two models. The critical value that must be exceeded to choose the more complex model can be interpreted as a penalty for the more complex model. However, this test can be applied only to nested models.

Two other general indices for comparing models, whether they are nested or not, are the *Bayesian information criterion (BIC)* and the *Akaike information criterion (AIC)*. The *BIC* is a Bayesian method. For very large sample sizes, the model with the lowest *BIC* corresponds to the model with the highest Bayes factor. The *BIC* is computed from G^2 by *adding* a penalty for the number of model parameters:

$$BIC(\text{model M}) = G^2(\text{model M}) + \ln(N) \cdot p_M,$$

where N equals the total number of observations and p_M is the number of model parameters. *AIC* is based on information theory. For very large sample sizes, the model with the lowest *AIC* corresponds to the model that produces a probability distribution with smallest discrepancy (according to a Kullback-Leibler discrepancy) between the model and the true distribution. *AIC* is computed from G^2 by adding a different penalty for number of parameters:

$$AIC(\text{model M}) = G^2(\text{model M}) + 2 \cdot p_M,$$

where p_M is the number of model parameters. The *AIC* and *BIC* indices provide a trade-off between accuracy and parsimony for model selection.

Another method for evaluating models is based on cross-validation. The basic idea of cross-validation is to estimate the parameters from one subsample of all the data set (called the *calibration* or *training data set*), and then use these same parameters to make predictions for the holdout subsample (the *test data set*). A special example of this method is "leave one out" cross-validation, in which the training set contains all but one observation, and the procedure is repeated, with each observation taking a turn as the observation that is left out. The model that predicts best for the test data set (using some lack of fit criterion, such as G^2) is then chosen as the best model.

Next, we present three ways to estimate the parameters and compare the three models discussed here using the data from Busemeyer, Matthews, and

Wang (2006). The first method, called the aggregate method, is to ignore individual differences (i.e., assume homogeneity of parameters across individuals so that $\theta_i = \theta$ for all i) and to fit the data from all eighteen trials and from all eighty-eight participants (fit all $N = 18 \cdot 88 = 1{,}584$ observations). This is not such a good idea if there are large individual differences, though, because the model parameters change across individuals, and averaging across individual differences can produce a misleading comparison of models (Estes & Maddox, 2005).

The second method, called the individual method, is to fit the eighteen choices from each individual, which allows a separate parameter vector θ_i to be estimated from each person. This method requires estimating a very large number of parameters—one set for each of the eighty-eight participants. This is not such a good idea if there is only a little data from each person, though, because the parameter estimates will be very noisy.

The third method is to use hierarchical modeling. On the one hand, it is like the aggregate method, which uses all eighteen trials from all eighty-eight participants to estimate the parameters. But on the other hand, it is like the individual method, which allows the parameters to vary across individuals by assuming a distribution over the parameters (θ_i is assumed to have some distribution, such as a normal distribution). The problem with this method is with formulating an appropriate distribution of parameters over individuals. First, we present the aggregate method.

4.4 Aggregate Method

In this case, we assume that there are no individual differences. This is not a good assumption, but it provides a starting point for analyses, and we can compare these results with the other methods. In this case, we estimate only a few parameters from all $N = 88 \times 18$ observations. Using this method, the model parameters were estimated by finding the parameter vector θ that maximized the total log likelihood (*TLL*). In general, the models are nonlinear, so we used a mathematical programming language like Python, R, MATLAB, Mathematica, or others. We will use MATLAB in this chapter. (See the website for all the programs used https://github.com/jbusemey/ChoiceModeling.git, and see the appendix at the end of this chapter for a simple example program.) The programs are used to search and find the maximum likelihood estimates.

Table 4.3
Parameter estimates and G^2 for each model.

	w	a	B	c	d	G^2
Model A	0.5295	0.3146	0	1	0	1,598.8
Model B	0.00001	.1398	1.3917	1	0	1,462.9
Model C	0.6017	0.5228	0	1	.1425	1,594.2
Model A+	0.00001	1.1977	0	.6367	0	1,538.5
Model B+	0.5159	0.5246	1.3887	.5960	0	1,449.4
Model C+	0.00001	1.2459	0	.6971	.0772	1,536.6

Parameters equal to 0 or 1 were fixed and not free. *Model+* indicates that the utility parameter c is free.

Table 4.3 shows the maximum likelihood parameter estimates and the G^2 (badness of fit) for each model. For the first three rows, the utility parameter was fixed to $c=1$, a risk-neutral utility function; and in the last three rows, it was allowed to be a freely estimated parameter. The penalty for each additional parameter used to determine *BIC* equals $ln(88 \cdot 18) = ln(1584) = 7.37$. The penalty for each additional parameter for the *AIC* is simply 2. The chi-square value that needs to be exceeded for rejecting a simpler model that has one less parameter than another model equals 3.84. In this case, the *BIC* is most conservative (i.e., it favors simple models), and the *AIC* is least conservative (i.e., favors more complex models).

As can be seen, there is little difference in G^2 badness of fit between Models A and C, but Model B is much lower (and better) than both. Furthermore, the difference favoring Model B exceeds Model A by more than the penalty 7.37 for *BIC*, and there is no penalty difference for Models B and C because they have the same number of parameters. Models including a utility parameter generally do better than models without the utility parameter. For example, the change produced by adding the utility parameter to Model B equals 13.1, which exceeds the *BIC* penalty, 7.37.

Note that the parameter estimates dramatically change across models. For example, Model B places almost equal weight on the probability that the opponent will defect when the utility parameter is included, but Model B puts all the weight on the probability that the opponent will cooperate when the utility is fixed to neutral, $c=0$. This difference shows that interpretation of parameters, such as the belief that an opponent will defect, can depend heavily on the model.

Table 4.4
Predictions from Model B+ obtained by using the parameter estimates for aggregate data.

	P(D|unknown)	P(D|D)	P(D|C)
Matrix 1	0.64 (.66)	0.89 (.93)	0.86 (.82)
Matrix 2	0.72 (.67)	0.94 (.91)	0.87 (.83)
Matrix 3	0.62 (.64)	0.88 (.92)	0.86 (.84)
Matrix 4	0.62 (.59)	0.88 (.89)	0.86 (.86)
Matrix 5	0.73 (.78)	0.94 (.94)	0.87 (.85)
Matrix 6	0.57 (.61)	0.84 (.90)	0.85 (.84)
Mean	0.65 (.66)	0.89 (.92)	0.86 (.84)

Observed results are shown in parentheses.

The predictions of Model B+ are shown in table 4.4. As can be seen, this model does a fairly good job of fitting the average results.

4.5 Individual Method

Now the models are fit to the eighteen observations from each person separately, producing eighty-eight sets of parameters θ_i for i = 1, 2, ..., 88 and producing one set of parameters for each person. The estimates in this case are obtained by maximizing the likelihood LL_i for person *i*. Table 4.5 shows total G^2, summed across the individual G_i^2 values, which is smaller than total G^2 based on the aggregate method because we are fitting many more parameters. The *BIC* penalty for an extra parameter for each model now equals 88·ln(18) = 254.3527, the *AIC* penalty equals 2·88 = 176, and the chi-square value for eighty-eight degrees of freedom equals 110.9. In this case, the *BIC* is most conservative, and the chi-square is least conservative.

The comparison of model fits obtained from fitting the individuals agrees fairly well with the fits obtained from the aggregate in this example. The first three columns show the total G^2 (summed across individual G^2 values) for the three models, assuming a risk-neutral utility function ($c = 1$), and the last three columns allow the utility parameter to be free. Model B once again is the best-fitting model. The improvement of Model B over Model A does not quite exceed the conservative *BIC* criterion, but it does exceed the *AIC* criterion and the chi-square value with a statistically significant difference. Model B and Model C have the same number of parameters, so there

Table 4.5
Total G^2 obtained by summing G_i^2 across individuals for each model.

	Model A $c=0$	Model B $c=0$	Model C $c=0$	Model A+ c free	Model B+ c free	Model C+ c free
Total G^2	1,265.6	1,014.7	1,178.6	1,084.8	893.4	953.3

is no extra penalty when comparing these two, but Model B provides a better fit than Model C.

The parameters obtained from the individual fits show a much different picture than those obtained from the aggregate analysis. Figure 4.1 shows the frequency distribution of the parameter values for Model B (including the utility parameter). Notice that the maximum likelihood fits to individuals tend to push the estimates toward either extreme to produce bimodal distributions. This problem stems from fitting individuals with a small amount of data from each person.

Model B (including the utility) seems to be the best model when using either aggregate or individual analysis using maximum likelihood estimation methods. To see how well this model predicts the results, we computed the predictions for each individual using the individual parameter estimates, and then averaged the predictions across individuals. These predictions are shown in table 4.6 (for comparison, observed proportions from table 4.2 are shown in parentheses).

As can be seen by comparing tables 4.5 and 4.2, Model B (with utility) fits less well to the average data than the direct fit to the average data: the predictions systematically overestimate for the unknown case.

4.6 Hierarchical Bayesian Method

The last method discussed estimated the model parameters using all eighteen choices from all eighty-eight participants by assuming that the parameters from each participant are sampled from a specified distribution, such as a normal distribution. We will use a different estimation method for this case. Rather than using a maximum likelihood estimation method, we will use a Bayesian estimation method. Maximum likelihood provides a single point estimate for each parameter vector that maximizes the likelihood of the data. The Bayesian method provides a posterior joint distribution

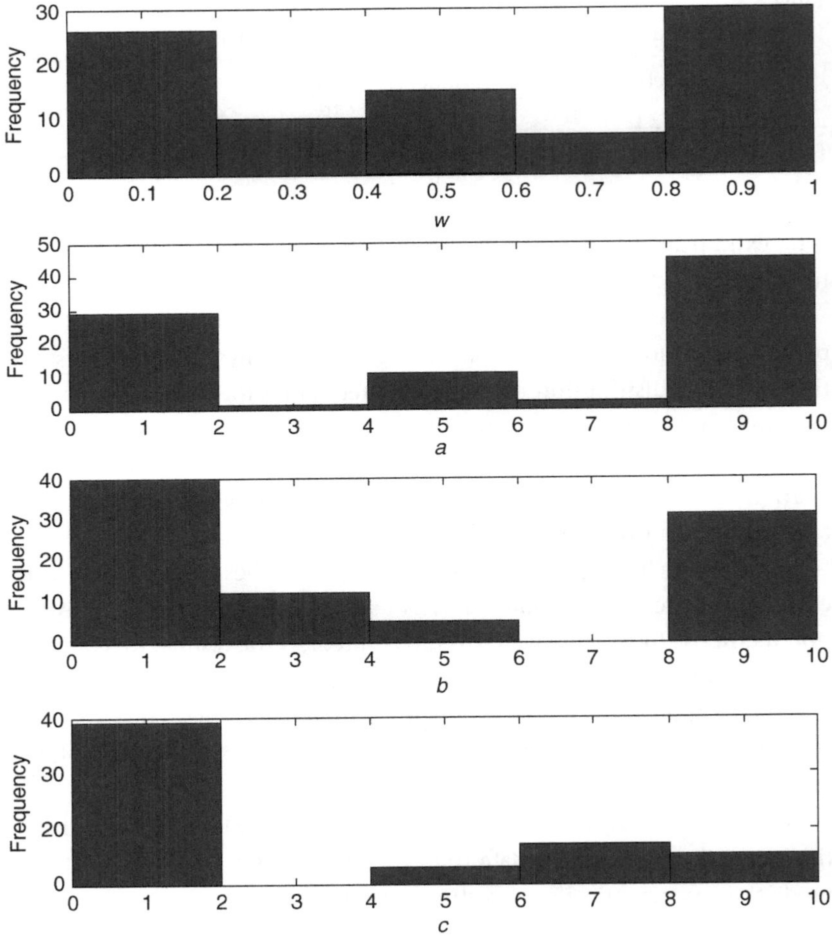

Figure 4.1
Frequency distribution of parameters for Model B obtained by fitting individuals.

over the parameters conditioned on the observed data (see Gelman et al., 2016; Kruschke, 2015; Lee & Wagenmakers, 2014).

To describe the Bayesian hierarchical estimation method, it will be helpful to define some additional notation. Earlier, we defined O_i as the vector of eighteen observed choices obtained from person *i*. The super-data vector, containing all eighty-eight individual data vectors, is denoted as $O = [O_1, O_2, \ldots, O_{88}]$. We also defined θ_i as the vector of parameters for the *i*th person

Table 4.6
Predictions obtained by using the parameter estimates for each individual and then averaging across individuals.

	P(D|unknown)	P(D|D)	P(D|C)
Matrix 1	0.73 (.66)	0.89 (.93)	0.88 (.82)
Matrix 2	0.79 (.67)	0.94 (.91)	0.89 (.83)
Matrix 3	0.73 (.64)	0.89 (.92)	0.88 (.84)
Matrix 4	0.70 (.59)	0.88 (.89)	0.88 (.86)
Matrix 5	0.79 (.78)	0.94 (.94)	0.89 (.85)
Matrix 6	0.65 (.61)	0.87 (.90)	0.87 (.84)
Mean	0.73 (.66)	0.90 (.92)	0.88 (.84)

Observed proportions are shown in parentheses.

when using a particular model. The super-parameter vector, containing all eighty-eight individual parameter vectors, is denoted as $\theta = [\theta_1, \theta_2, \ldots, \theta_{88}]$.

The basic idea of a hierarchical model is that we first sample a parameter vector for each individual from a parent distribution, and then we sample predictions for the individual given the person's parameter vector. Therefore, when using a hierarchical model, we assume that the super-parameter vector θ is generated from what is called a *parent density distribution,* denoted as $f_M(\theta)$.

In particular, we assumed that the parameter a_i for person i for the logistic model was sampled from a normal distribution with mean μ and a standard deviation equal to σ. In general, the logistic coefficient can range from positive to negative infinity, although it is expected to be positive. A normal distribution is commonly used for distributions that range from positive to negative infinity, and it is also commonly used for the logistic coefficient, as we have done here. For similar reasons, we also assumed that the bias parameter b of Model B and the weight parameter d of Model C were sampled from normal distributions.

We assumed that parameter w_i for each person i was sampled from a Beta distribution, $B(\alpha, \beta)$, which is commonly used when we need to sample a parameter that is continuously distributed between 0 and 1. The Beta distribution is defined by two parameters (α, β) that provide the distribution with a great deal of flexibility. The mean of the Beta distribution equals $\alpha/(\alpha+\beta)$, and the variance equals $\alpha \cdot \beta/[(\alpha+\beta)^2(\alpha+\beta+1)]$. Figure 4.2 illustrates the probability density of the Beta distribution for a selection of (α, β) that is typical of the distribution used in our analyses.

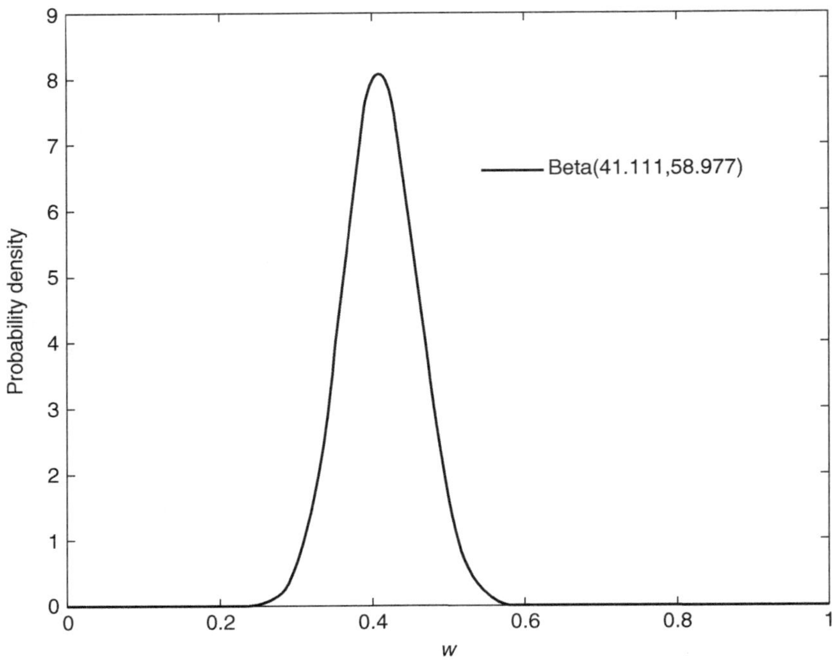

Figure 4.2
Beta distribution using typical parameters for the application used in our analyses.

We assumed that the utility parameter c_i was sampled from a gamma distribution Gamma(α, β). The utility parameter, c_i, must be nonnegative, but it is not necessarily bounded from above. The gamma distribution is commonly used for this type of distribution (bounded below by zero). Like the Beta, it has two parameters that provide the distribution a great deal of flexibility. The mean of the gamma distribution equals α/β and the variance equals α/β^2.

Therefore, parent density f_M is composed of the densities (normal, Beta, and gamma) described earlier in this chapter. The parent distribution has parameters that we call *hyperparameters*. We can collect all of the hyperparameters of the parent distribution (μ and σ for a and b, and α and β for c and d) into one vector of hyperparameters, denoted as ϕ. We assume that the individual parameters sampled for one participant from the parent distribution are dependent only on the hyperparameters, and they are conditionally independent of the other participants:

$$f_M(\theta \mid \phi) = f_M(\theta_1 \mid \phi) \cdot f_M(\theta_2 \mid \phi) \cdots f_M(\theta_{88} \mid \phi)$$

For the hierarchical model, we also assume that the data obtained from an individual depend only on the parameter vector θ_i for that individual and are independent of other individuals such that the probability of the data from all eighty-eight participants, given the parameters from all participants, equals the product of likelihoods from individuals:

$$p_M(O \mid \theta) = p_M(O_1 \mid \theta_1) \cdot p_M(O_2 \mid \theta_2) \cdots p_M(O_{88} \mid \theta_{88}).$$

The problem for the hierarchical Bayesian method is to estimate the posterior density over (ϕ,θ), given the observed data O for a particular model M, which is denoted as $p_M(\phi,\theta \mid O)$. Recall that according to the Bayes rule, if H is our hypothesis and D is our data, then $p(H \mid D) = p(H) \cdot p(D \mid H) / p(D)$, where $p(H)$ is our prior probability and $p(H \mid D)$ is our posterior probability given the data. In this application, the parameter vector (ϕ,θ) plays the role of the hypothesis H, and the observation vector O plays the role of the data D, and

$$p_M(\phi,\theta \mid O) = \frac{p_M(\phi,\theta) \cdot p_M(O \mid \phi,\theta)}{p_M(O)}.$$

gives the posterior probability density of the parameters given the data. The prior probability density is defined by the product $p_M(\phi,\theta) = p_M(\phi) \cdot f_M(\theta \mid \phi)$. We still need to specify this prior, but using this prior and the likelihood of the data given the parameters, we can use the Bayes theorem to obtain our joint posterior density given the data $p_M(\phi,\theta \mid O)$. Furthermore, we can obtain marginal probability densities such as $p_M(\phi \mid O)$ by integrating over the other parameters:

$$p_M(\phi \mid O) \propto \int p_M(\phi,\theta) \cdot p_M(O \mid \theta) d\theta.$$

Finally, we can compute the mean and standard deviations or the percentiles (such as the median) of a parameter by using the marginal distribution of the parameter.

Steps in Bayesian estimation The first step in this analysis is to specify the prior distributions for each parameter. We wish to choose a fairly broad prior, but also one that does not put much weight on implausible values. For the mean μ of the normally distributed parameters (mean of a, b, and d), we chose a normal prior with mean equal to 0 and standard deviation equal to 10 [i.e., $\mu \sim N(0,10)$]. For the standard deviation of the normally distributed parameters (std of a, b, d), we chose a uniform distribution $\sigma \sim [.01,10]$. For the Beta distribution, we chose uniform priors $\alpha \sim [1,100]$ and $\beta \sim [1,100]$. Finally, for the gamma distribution, we also chose uniform priors $\alpha \sim [1,100]$ and $\beta \sim [1,100]$. We also assume independence among the hyperparameters so that the joint prior probability is a product of these marginals.

The next step is to determine the posterior distribution using the data and the priors. Usually it is not possible to derive an analytical solution to the posterior distribution. Instead, it is common to use a Markov chain Monte Carlo (MCMC) sampling algorithm that samples points in the parameter space based on an importance sampling algorithm. The MCMC algorithm starts from an initial state and then conducts a kind of random walk through the parameter space, guided by the likelihood of a parameter vector given the data. After some initial searching, the process converges on an equilibrium distribution that is proportional to the posterior distribution of interest.

Different MCMC algorithms are commonly available on open-source websites and can be downloaded for this purpose. We used the Stan program, developed by the Stan development team in 2015 (http://mc-stan.org). In particular, we used the MatlabStan version developed by Brian Lau (2015). The code that we used is provided on our website (https://github.com/jbusemey/ChoiceModeling.git). Finally, the MCMC algorithm will provide a large sample of parameter vectors from the posterior density, from which we can construct histograms of the marginal distributions of the parameters; compute estimates of means, standard deviations, and medians of the marginal distributions; and compile many other statistics.

When using an MCMC algorithm, a number of algorithm selections need to be determined. The first is called the *burn-in period*, which is the number of initial samples that are eventually discarded because the MCMC algorithm has not converged on the equilibrium distribution. We used 5,000 samples for our burn-in period, which proved to be sufficient. After the burn-in period, we sampled an additional 10,000 samples for use in computing the estimates of the posterior distributions.

To check whether the MCMC algorithm did in fact converge after the burn-in period, it is common for the computer package to compute a statistic called *R* from the samples for each parameter. The R statistic requires running several simulations to check consistency across simulations. Essentially, R is a comparison of the variance between simulations to the variance within a simulation. If the algorithm has converged, then the R statistic should be close to 1.0; Gelman et al. (2016) recommends that R statistics greater than 1.1 are not acceptable. We ran four separate simulations, and the results produced R statistics very close to 1, with all the Rs less than 1.003. Given the consistency across the four simulations, we combined the four simulations to form one large sample with 10,000 samples.

The third step is to examine the marginal distributions of the parameters generated by the MCMC algorithm. The program provides a large matrix containing M = 10,000 rows, with one row for each simulation, and p columns, with one column for each parameter. Selecting a single column picks out the samples for one parameter. Using this column of sample values, we can compute a histogram of relative frequencies—or better yet, we can filter the histogram to produce a smoothed estimate of the posterior probability density.

Now we will focus on the parameters used in the parent distribution. First, we transformed the two parameters for the Beta distribution that determine the distribution of probability weights by computing the ratio $w = \alpha/(\alpha + \beta)$ for each sample (the mean of the Beta). Next, we transformed the two parameters for the gamma distribution that determined the distribution of the utility parameter by computing the ratio $c = (\alpha/\beta)$ for each sample (the mean of the Gamma). Then we plotted the relative frequencies across all the samples of the values of each parameter for Model B (including the utility parameter). Figure 4.3 illustrates the marginal distributions of the means of the parent distribution for the main four parameters (w = probability weight, a = logistic slope, b = bias, and c = utility parameter). The smooth curve represents the density estimate, the vertical line represents the median, and the small raised interval represents the .95% highest-density interval containing 95% of the distribution. The posterior means and standard deviations of these distributions are shown in table 4.7.

One can see that the Bayesian method no longer produces the bimodal distributions; instead, it provides unimodal distributions around values that are consistent with past theory and research. In particular, the mean slope of the logistic is near 1, and the mean of the utility parameter is risk averse. The mean probability weight given to the opponent choosing to defect is .25.

The posterior predicted means for Model B (c free) are shown in table 4.8. These were obtained by using the 10,000 samples to compute the posterior mean prediction for each person, and then averaging across participants.

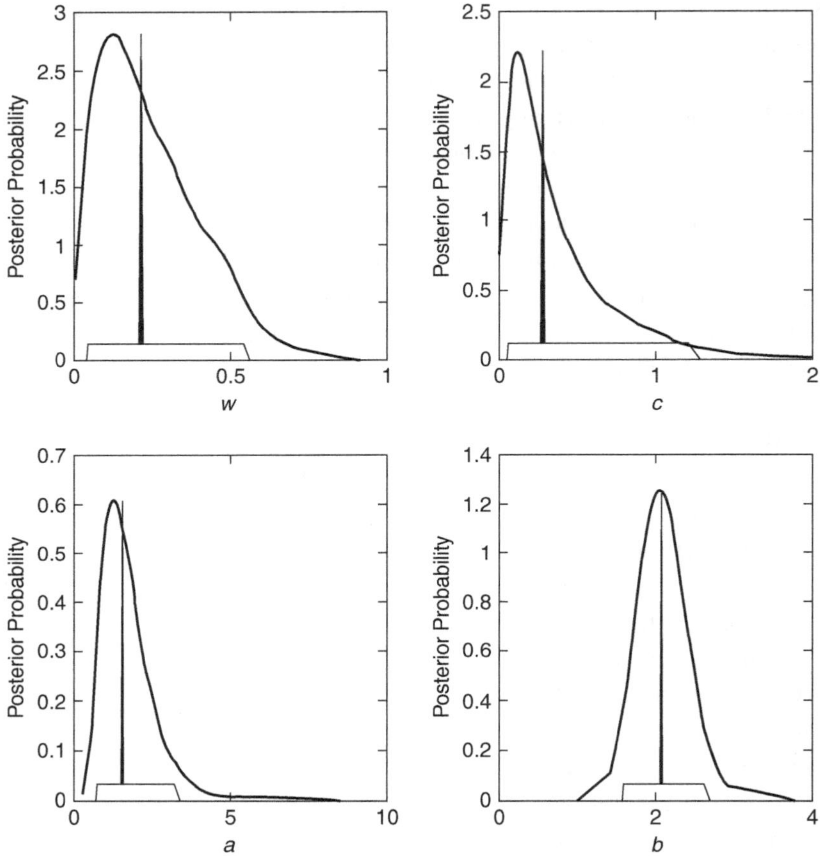

Figure 4.3
Posterior marginal distribution of the means for the hyperparameters (*w*, *a*, *b*, and *c*) used in the parent distribution.

The predictions generated by the Bayesian estimation method are more accurate for the average results than those produced by averaging the individual estimates.

The fourth step is to compare the competing models using the posterior distributions. There are various ways to compare models using Bayesian estimation methods. One important method is to use cross-validation. In particular, we will use the Bayesian "leave one out" cross-validation criterion to compare models—specifically, the estimate of the leave one out expected

Table 4.7
Posterior means and standard deviations of hyperparameters from parent distributions.

	w	*a*	*b*	*c*
Mean	0.2466	1.7036	2.0797	0.4197
Standard deviation	0.1597	0.8207	0.3200	0.5046

Table 4.8
Posterior mean predictions from Model B (*c* free).

	P(D|unknown)	P(D | D)	P(D | C)
Matrix 1	0.66 (.66)	0.89 (.93)	0.86 (.82)
Matrix 2	0.73 (.67)	0.92 (.91)	0.87 (.83)
Matrix 3	0.65 (.64)	0.88 (.92)	0.86 (.84)
Matrix 4	0.65 (.59)	0.88 (.89)	0.86 (.86)
Matrix 5	0.74 (.78)	0.92 (.94)	0.88 (.85)
Matrix 6	0.63 (.61)	0.86 (.90)	0.86 (.84)
Mean	0.68 (.66)	0.89 (.92)	0.87 (.84)

Predictions obtained by computing the posterior mean for each person and then averaging across participants using the 10,000 samples of parameters from Model B (*c* free). Observed proportions shown in parantheses.

log pointwise predictive density, denoted as elpd (see Vehtari, Gelman, & Gabry, 2016). Model A, Model B (with neutral risk aversion setting $c=1$), and Model B (with a free parameter for risk aversion). Technically, we compute the following estimation of leave one out:

$$est\ LOO = \sum_{n=1}^{88} \ln(p_M(O_i \mid O_j, j \neq i)),$$

which is the sum of the log posterior predicted probability of the next observation, given all of the previous observations. Leave one out is a *goodness* of fit index, and we select the model with the *highest* leave one out value. The Stan program provides an algorithm for computing this estimate. Using this algorithm, we obtain the following Bayesian leave one out estimates:

$elpd = -19.784$ for Model A,

$elpd = -13.7317$ for Model B ($c=1$),

$elpd = -19.7294$ for Model B (c free)

$elpd = -20.2119$ for Model C ($c=1$)

The difference between the leave one outs for Model A and Model B ($c=1$) equals 3.56, and the standard error of this leave one out difference equals 2.86. We conclude that Model B ($c=1$) is the best model for leave one out cross-validation. The choice of Model B ($c=1$) over Model A and Model C also agrees with the aggregate and individual analyses described earlier.

4.7 Summary

In this chapter, we reviewed three methods for estimating and comparing predictions of choice models: one based on the aggregate method, which ignores individual differences; the second based on fitting the model separately to each individual, which requires estimation of a much larger number of parameters with little data; and the third, the hierarchical method, which uses all the data from all the participants by building a parent distribution for generating the individual differences. We also reviewed two major methods for estimating model parameters: one based on maximum likelihood estimation that provides a single point estimate, and another based on Bayesian estimation, which provides a posterior distribution over the parameters. Finally, we considered several methods for comparing models, including those based on maximum likelihood estimation (such as chi-square tests for nested models), and *AIC* and *BIC* model indices; and we also presented an example of cross-validation using the Bayesian estimation method.

Of course, this is just a brief introduction to these topics, and the reader should consult reference books that go much deeper into these topics, such as Busemeyer and Diederich (2010), Lewandowsky and Farrell (2011), and Busemeyer, Wang, Townsend, and Eidels (2015).

4.A Appendix

This appendix shows example computer code used to generate the predictions from Model B with fixed utility equal to $c=1$. This can be obtained from the web site https://github.com/jbusemey/ChoiceModeling.git.

The *main program* first reads the required data matrices (lines 6 and 26), and then loops through the subjects (lines 44–69) and calls the search program (line 56). The search program calls the function FitMerv3 and passes a vector of three parameters, w, a, and b, in the variable called Parm to the function FitMerv3. Then FitMerv3 evaluates the G^2 fit produced by that parameter vector and returns this fit to the search program in the output variable. The search program repeatedly tries different parameters to find values that reduce G^2, and the search process continues until the reduction in G^2 falls below a tolerance criterion. After reaching the tolerance level, the search program returns the minimum G^2 and the parameter vector that produced this minimum to the main program. The main program stores and prints the results for the researcher (lines 63–64). The *Generate predictions* program (line 71) is used to compute the predictions of the individuals after estimating the parameter from the Main program.

The function FitMerv3 (line 96) takes the input parameters stored in Parm selected by the search algorithm, as well as three additional variables provided by the main program. The variable PS is a data matrix that indicates the type bonus trial (in column 7), the opponent's choice (in column 6), the type of game (in columns 3–10), and the observed choice (in column 4) on a particular trial (row). The variable PM is a data matrix built by the programmer that contains information about the payoffs for a game.

The main part of the function FitMerv3 contains a *for … end* loop. This loops through the trials (rows) to compute the prediction on each trial. The first three lines determines what type of weight should be applied on each trial. The next two lines determine the payoff for each trial. The next two lines compute the predicted choice probability for a trial (*eps* is used to avoid predicting exactly 1 or zero, which causes problems for computing the Chi, which requires taking the log of the probability). The last three lines compute the Chi lack of fit statistic.

```
% Main Program
clc
clear

load Merv1
% PD  88 subj, 40 trials per subj
% 40 trials, 28 no bonus , 12 bonus
% N=40*88=3520, 28*88=2464
% N x 10 data matrix
%  1    2   3      4     5        6        7     8
% subj trial matrix choice opponent oppchoice bonus payoff
 9   10
RT  Gnumb
 ns=88; % no subj for individual fits
```

```
% [By c]=find(PD(:,7)==2);  % bonus trials
% [Bn c]=find(PD(:,7)==1);  % not bonus
[G1, c]=find(PD(:,10)==1); % PD only   18 trials, 12 bonus 6
no bonus
PD=PD(G1,:);


PD(:,4)=2 - PD(:,4);   % 1=defect, 0=coop
N=size(PD,1);     % 18*88=1584 only PD, 28*88=2464 no bonus,
or 40*88=3520 all
nt=N/ns;    % no trials per subj  18 PD only or 28 (no bonus)
or 40 (all)

load Matrices    % payoff matrices used in games
% PMall 6 matrices x 8 cols x 5 games
% 1=PD,2=CH,3=IS,4=BS,5=FR : types of games
% cols
% M Y    M Y    M Y    M Y
% D D    D C    C D    C C
% 1 1    1 2    2 1    2 2    num code for D,C
% col's contain payoffs for a game

np=3;   % no parms: wgt per game (6), choice parm (1), bonus
effect (1)
ParmV=zeros(ns,np);
ChiV=zeros(ns,1);

nr=3;   % no of random start parms
lb=zeros(np,1);
% ub=[ones(5,1); 10.*ones(np-5,1)];
ub=[1 ; 10*ones(np-1,1)]';   % for PD model only


    for sub=1:ns
        sub
        RepParm=zeros(nr,np);
        RepChi=zeros(nr,1);
        for rep=1:nr
```

```
            % Parm0=randn(np,1);
            Parm0=rand(np,1);
            %  Parm0=[.4 .2 10]';
            indx=(sub-1).*nt+(1:nt);  % individual subject
            PS=PD(indx,:);
            options=optimset('MaxFunEvals',10000,'MaxIter',
              10000,…'TolX',1e-10);
            [param, Chi]=fmincon(@(Parm) FitMerv3(Parm, PS,
              PMall, nt),…
                Parm0, [], [], [], [], lb, ub, [], options);
            [rep Chi/1000]
            RepParm(rep,:)=param;
            RepChi(rep)=Chi;
        end % rep
         [mv mindx]=min(RepChi);
         ParmV(sub,:)=RepParm(mindx,:);
         ChiV(sub)=mv;
    end % subj

    %   save MFit3 ParmV ChiV

% Generate predictions
clc
clear

load MFit3
Comp=zeros(ns,1);
Pred=zeros(ns,6,3);
% six games, 2 bonus trials per game,
% 1 no bonus trial per game

    for sub=1:ns
        indx=(sub-1).*nt+(1:nt);
        PS=PD(indx,:);
        Parm=ParmV(sub,:);
```

```
        [Pr, Chi]=FitMerv3p(Parm,PS,PMall,nt);
        Pred(sub,:,:)=Pr;

    end

    disp(squeeze(mean(Pred,1)))  % for avg of indiv fit

function [Chi]=FitMerv3(Parm, PS, PM, nt)
% function [Chi]=FitMerv3(Parm, PS, PM, nt)
% three parm -- use only PD trials, bonus and not bonus trials

% W=1./(1+exp(-Parm(1:5)));    % include for unconstrained

W=Parm(1);
a=Parm(2);
b=Parm(3);
% b=0;
eps=1./(10.^5);

Chi=0;
B=PS(:,7); % bonus indicator, 1=not bonus, 2=bonus
Op=PS(:,6); % Opp choice

for row=1:nt
    w1=W; w2=1-w1;
    w1=(B(row)==1).*w1+(B(row)==2).*(Op(row)==1);
    w2=(B(row)==1).*w2+(B(row)==2).*(Op(row)==2);
    Pay=PM(PS(row,3),:,PS(row,10));
    Pay=(w1.*Pay(1)+w2.*Pay(3))-(w1.*Pay(5)+w2.*Pay(7));
    Pr=1./(1+exp(-(a.*Pay+(B(row)==2).*b)));
    Pr=eps+(1-2.*eps).*Pr;
    ch=PS(row,4); % obs choice
    LL=ch.*log(Pr)+(1-ch).*log(1-Pr);
    Chi=Chi+-2*LL;
end

end
```

```
function [PP, Chi]=FitMerv3p(Parm, PS, PM, nt)
% function [Chi]=FitMerv3p(Parm, PS, PM, nt)
% three parm -- use only PD trials, bonus and not bonus trials

% W=1./(1+exp(-Parm(1:5)));    % include for unconstrained

W=Parm(1);
a=Parm(2);
b=Parm(3);
% b=0;
eps=1./(10.^5);

Chi=0;
B=PS(:,7);
Op=PS(:,6);

% Pr1=0; Pr2=0; Ch1=0; Ch2=0;

PP=zeros(6,3);

for row=1:nt
    w1=W; w2=1-w1;
    w1=(B(row)==1).*w1+(B(row)==2).*(Op(row)==1);
    w2=(B(row)==1).*w2+(B(row)==2).*(Op(row)==2);
    Pay=PM(PS(row,3),:,PS(row,10));
    Pay=(w1.*Pay(1)+w2.*Pay(3))-(w1.*Pay(5)+w2.*Pay(7));
    Pr=1./(1+exp(-(a.*Pay+(B(row)==2).*b)));
    Pr=eps+(1-2.*eps).*Pr;
    ch=PS(row,4);
    LL=ch.*log(Pr)+(1-ch).*log(1-Pr);
    Chi=Chi+-2*LL;

%      Pr1=Pr1+(B(row)==1).*Pr;    % no bonus
%      Pr2=Pr2+(B(row)==2).*Pr;    % bonus
%      Ch1=Ch1+(B(row)==1).*ch;
%      Ch2=Ch2+(B(row)==2).*ch;
     PP(PS(row,3),2)=PP(PS(row,3),2)+(B(row)==2).*(Op(row)==1).*Pr;
     PP(PS(row,3),3)=PP(PS(row,3),3)+(B(row)==2).*(Op(row)==2).*Pr;
```

```
    PP(PS(row,3),1)=PP(PS(row,3),1)+(B(row)==1).*Pr;
%       [PS(row,3) B(row) Op(row)]
          PP(PS(row,3),:)

end

% Pr1=Pr1/6;
% Pr2=Pr2/12;
% Ch1=Ch1/6;
% Ch2=Ch2/12;
% Pr=[Pr1 Ch1 Pr2 Ch2];

end
```

5 Multidimensional Signal Detection Theory

Starting in the 1950s, psychologists started developing their own theories of decision-making for cognitive tasks such as sensation, perception, memory recognition, and categorization (Tanner & Swets, 1954). The earliest and most prominent was *signal detection theory* (Swets, Tanner, & Birdsall, 1961), later popularized by Green and Swets (1966). The purpose of signal detection theory is to describe how people make decisions when an uncertain state of nature is inferred from a noisy sample of data generated by that state.

The first application of this model was back in World War II, such as detecting whether a sonar signal arises from an enemy submarine. Later, other applications occurred, such as medical (e.g., deciding whether an x-ray image from a patient shows a benign or cancerous tumor), or criminal justice (e.g., deciding whether a suspect is the actual criminal from a lineup) decisions, and many others. The signal detection model has been used to predict both accuracy (the probability of taking the appropriate action for a given state of nature) and the confidence rating for the action that was chosen. This chapter presents some of the basic ideas, but there are many good books that provide more details (e.g., Green & Swets, 1966; Wickens, 2002; Macmillan & Creelman, 2004).

5.1 The Basic Signal Detection Task

The signal detection task is closely related to categorization tasks, with payoffs for correct or incorrect responses. In a basic signal detection task, a decision-maker (e.g., doctor, military operator, or witness) chooses one action from a set of K actions (e.g., a radiologist must decide whether disease A is present or absent, and in this case, $K=2$). The payoffs produced by choosing an action depend on an unknown state of nature, and in general, there could be M possible states of nature (e.g., the true state s of a patient might be that disease A is present, or the true state might be n, with no disease present, and in this case, $M=2$). The decision-maker is provided with some data X sampled from the existing (but yet unknown) state of nature, which provides information about the current state of nature. The task is to choose an action from the set of K actions after inferring the state of nature from a set of M possible states on the basis of sample data (X).

The payoff matrix for this example, which is the simplest case with $K=M=2$, is shown in table 5.1. A hit occurs when the state is s (disease state is present) and the agent chooses action A (diagnose disease). A false alarm occurs when the state is n (no disease state) and the agent chooses action A (diagnose disease). A miss occurs when the state is s (disease is present) and the agent chooses action B (decides that disease is absent). A correct rejection occurs when the state is n (no disease state) and the agent chooses action B (decides that disease is absent).

The agent is provided some data or sample information concerning the state of nature. In general, this is represented by an r dimensional vector X, such as a multidimensional representation of an x-ray, ultrasound, or magnetic resonance image (MRI). Presumably, the agent knows something about the likelihood of the data X given each state of nature (e.g., the radiologist has a great deal of past experience with feedback). Then the data can be evaluated by comparing the likelihood of the data $p(X|s)$ when the signal state s is present with the likelihood of the data $p(X|n)$ when the no signal n occurs. The log likelihood ratio $\ln L(X) = \ln p(X|s) - \ln p(X|n)]$ provides a measure of the strength of evidence favoring signal s over no signal n provided by the data X when the true state is unknown.

Usually, a signal detection study includes a manipulation of two different conditions that use two different levels of signal strength: one condition with a signal present (state s, signal embedded in noise), and a second

Table 5.1
Payoff matrix for two actions and two states of nature.

	s: Disease state is present	n: Disease state absent
A: Diagnose disease	u_{As} = Hit (gain)	u_{An} = False alarm
B: Diagnose no disease	u_{Bs} = Miss (loss)	u_{Bs} = Correct rejection

Table 5.2
Hit and false alarm rates produced by a typical signal detection task.

	State *s*, signal present	State *n*, noise
Choose *A*	$\Pr(A \mid s)=$hit rate	$\Pr(A \mid n)=$false alarm rate
Choose *B*	$\Pr(B \mid s)=$miss rate	$\Pr(B \mid s)=$correct reject rate

condition with no signal (state *n*, a pure noise condition). The results produce a 2×2 table of choice proportions, each one conditioned on the state, as shown in table 5.2. For example, $\Pr(A \mid s)$ represents the proportion of trials in which the decision-maker chose *A*, given that a signal was presented, and $\Pr(B \mid s)=1-\Pr(A \mid s)$ is the proportion of misses.

5.2 The Optimal Decision Rule for the Two Action–Two State Signal Detection Task

Signal detection theory is a very general theory that is not restricted to any assumptions about optimality. However, it is often informative to derive the optimal rule from assumptions about the signal detection task and then compare human performance to this optimal rule. A decision-maker could have various types of objectives, but we will assume that the objective is to maximize expected utility. Under this assumption, the optimal solution to the signal detection/categorization problem combines Bayesian inference with the expected utility decision rule as follows.

The expected utility for each action equals

$$EU(A)=p(s \mid X)\cdot u_{As}+p(n \mid X)\cdot u_{An},$$

$$EU(B)=p(s \mid X)\cdot u_{Bs}+p(n \mid X)\cdot u_{Bn}.$$

The difference equals

$$EU(A)-EU(B)=p(s \mid X)\cdot(u_{As}-u_{Bs})-p(n \mid X)\cdot(u_{Bn}-u_{An}).$$

To derive the optimal rule, we will also need to use the Bayes theorem to infer state *s* from the data *X*. According to the Bayes rule,

$$p(s \mid X)=\frac{p(s\cap X)}{p(X)}=\frac{p(s)\cdot p(X \mid s)}{p(X)}.$$

Inserting the result from the Bayes rule, we obtain [ignoring $p(X)$ because this term eventually cancels out]

$$EU(A) - EU(B) = p(s) \cdot p(X \mid s) \cdot (u_{As} - u_{Bs}) - p(n) \cdot p(X \mid n) \cdot (u_{Bn} - u_{An}).$$

The optimal decision is obtained by choosing action A whenever $EU(A) - EU(B) > 0$, which implies that the agent should choose action A whenever

$$p(s) \cdot p(X \mid s) \cdot (u_{As} - u_{Bs}) > p(n) \cdot p(X \mid n) \cdot (u_{Bn} - u_{An}),$$

which in turn finally implies that A should be chosen whenever

$$L(X) = \frac{p(X \mid s)}{p(X \mid n)} > \frac{p(n)}{p(s)} \cdot \frac{u_{Bn} - u_{An}}{u_{As} - u_{Bs}}.$$

Taking the logarithms, we define the evidence for A over B as $Y = \ln L(X)$. Then the rule is to choose A whenever

$$Y = \ln L(X) > c, \text{ where } c = \ln\left(\frac{p(n)}{p(s)}\right) + \ln\left(\frac{u_{Bn} - u_{An}}{u_{As} - u_{Bs}}\right).$$

The evidence favoring the signal state s over the no signal or noise state n is measured by $Y = \ln L(X)$. Hereafter, we will use X to represent the r-dimensional data vector presented to the decision-maker, and Y will be used to represent the unidimensional evidence favoring state s over n extracted from the data.

The mean evidence, $E(Y \mid s)$, will be positive, favoring state s, when in fact state s is present; and $E(Y \mid n)$ will be negative, favoring state n, when in fact state n is present. In the standard model, which assumes a multivariate normal distribution of X, we obtain $E(Y \mid n) = -E(Y \mid s)$, and so $E(Y \mid s) - E(Y \mid n) = 2 \cdot E(Y \mid s)$. The magnitude of this mean determines the strength of the evidence provided by the data. However, performance on the task is also affected by the variance of the evidence, $V(Y \mid state)$. In general, this variance could depend on the state, but in most applications, it is assumed to be independent of the state. Low variance allows greater discriminability of the mean, and large variance decreases discriminability. The ratio, $d' = [E(Y \mid s) - E(Y \mid n)] / std(Y)$, is an index of discriminability.

The cutoff c is called the *bias parameter*. For example, if payoffs are symmetric (an equal cost for each type of error) but the priors favor state n (e.g., disease is rarely present or base rate is low), then the bias is positive, which means that the evidence has to be stronger than zero (strong evidence for signal) in order to choose A. Alternatively, if the priors are equal but the payoff ratio favors choosing action A (e.g., the gain for a hit is large and the loss for a false alarm is small), then the bias is negative, which means that

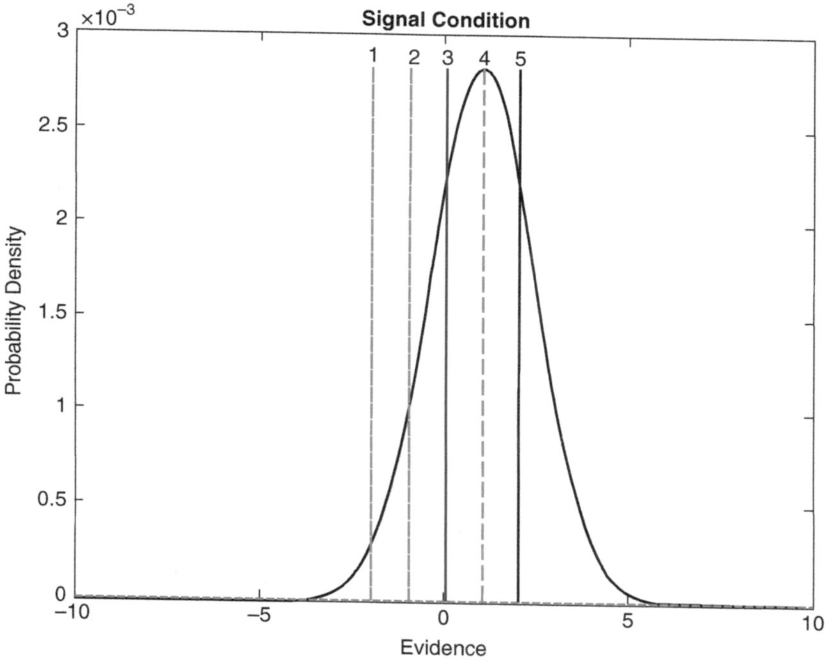

Figure 5.2
Five levels of bias ranging from low (1) to high (5).

area under the ROC. This nonparametric estimate of evidence strength is called the *A′ index of detectability* (MacMillan, 2002). Balakrishnan (1998) also provides nonparametric methods for estimating bias.

Assuming normal distribution with equal variance, we can derive simple parametric estimates of evidence strength and bias. In this simple case, the signal detection model reduces to a case V type of Thurstone random utility model. More formally,

$$p(Hit) = p(Y > c \mid s) = \frac{1}{\sigma\sqrt{2\pi}} \int_c^{\infty} e^{-\frac{1}{2}\left(\frac{y-\mu_s}{\sigma}\right)^2} dy$$

$$p(FA) = p(Y > c \mid n) = \frac{1}{\sigma\sqrt{2\pi}} \int_c^{\infty} e^{-\frac{1}{2}\left(\frac{y-\mu_n}{\sigma}\right)^2} dy.$$

Here, *p*(*Hit*) and *p*(*FA*) actually denote the population probabilities instead of the sample proportions.

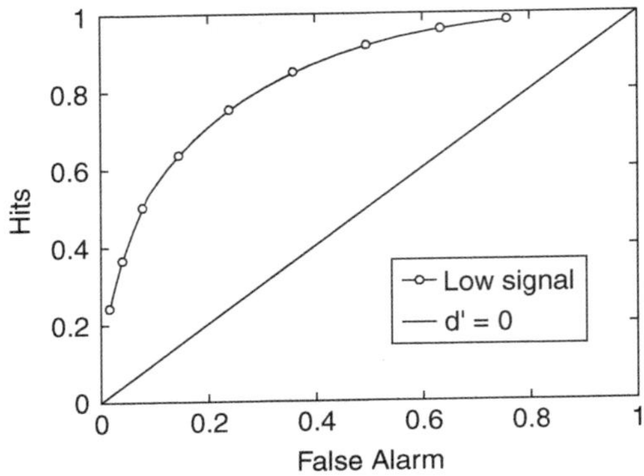

Figure 5.3
ROC curve. Probability of hits plotted as a function of the probability of false alarms, with a different point produced by each location of nine bias conditions.

It is convenient to transform the evidence scale Y to a Z scale using the definition $Z_i = \left(\frac{Y-\mu_i}{\sigma}\right)$. On this scale, the bias c on the Y scale is transformed to $\left(\frac{c-\mu_S}{\sigma}\right)$ on the Z scale under the signal state, and $\left(\frac{c-\mu_n}{\sigma}\right)$ on the Z scale under the no signal state. Then we can compute the probabilities using the transformed scores as follows:

$$p(Y > c \mid s) = p\left(Z_s > \left(\frac{c-\mu_S}{\sigma}\right)\right) = \frac{1}{\sqrt{2\pi}} \int_{\frac{c-\mu_s}{\sigma}}^{\infty} e^{-\frac{1}{2}z^2}\, dz = F\left(\frac{\mu_s - c}{\sigma}\right)$$

$$p(Y > c \mid n) = p\left(Z_n > \left(\frac{c-\mu_n}{\sigma}\right)\right) = \frac{1}{\sqrt{2\pi}} \int_{\frac{c-\mu_n}{\sigma}}^{\infty} e^{-\frac{1}{2}z^2}\, dz = F\left(\frac{\mu_n - c}{\sigma}\right),$$

where $F(z)$ is the standard (mean zero, unit variance) cumulative normal distribution function. The last equality on the right follows from the symmetry of the standard normal distribution: the area above the cutoff $\left(\frac{c-\mu_i}{\sigma}\right)$ equals the area below the cutoff $\left(\frac{\mu_i - c}{\sigma}\right)$. For convenience, we define $z_H = \left(\frac{\mu_S - c}{\sigma}\right)$ and $z_{FA} = \left(\frac{\mu_n - c}{\sigma}\right)$.

Using the inverse cumulative normal transformation, we can convert the estimated probabilities of hits and false alarms, Pr(Hit) and Pr(FA)), into z-values (but note that experimentally, we are observing only sample estimates):

$$z_H = \left(\frac{\mu_s - c}{\sigma}\right) = F^{-1}(\Pr(Hit))$$

$$z_{FA} = \left(\frac{\mu_n - c}{\sigma}\right) = F^{-1}(\Pr(FA)).$$

Using these two equations, we can solve for

$$d' = \left(\frac{\mu_s - \mu_n}{\sigma}\right) = (z_H - z_{FA})$$

$$b = \frac{c}{\sigma} + \frac{\mu_s + \mu_n}{2\sigma} = -\left(\frac{z_H + z_{FA}}{2\sigma}\right).$$

Recall that when the evidence is represented by the log likelihood, we obtain $\mu_s + \mu_n = 0$, in which case $d' = \frac{\mu_s - \mu_n}{\sigma} = \frac{2\mu_s}{\sigma}$ and $b = \frac{c}{\sigma}$.

At this point, we could use these simple formulas for computing d' and b from z_H and z_{FA} separately for each bias condition. But this is *not* the best way to estimate these parameters from a rigorous statistical point of view. A better method is to estimate these parameters from all the experimental data using the maximum likelihood (Dorfman & Alf, 1969) or Bayesian (Lee, 2008) methods discussed in chapter 4. The code for hierarchical Bayesian modeling of signal detection data is located at https://faculty.sites.uci.edu/mdlee/bgm/, under the computer link "Code for book."

The signal detection model provides a very strong *a priori* test. According to the theory developed in this discussion,

$$z_H = \left(\frac{\mu_s - c}{\sigma}\right) = \left(\frac{\mu_n - c}{\sigma}\right) + \left(\frac{\mu_s - \mu_n}{\sigma}\right) = z_{FA} + d'.$$

This implies that z_H is linearly related to z_{FA}, with unit slope and intercept d'. Suppose that we manipulate the bias across several levels, holding the strength of evidence constant. Then we can plot points of observed pairs (z_{FA}, z_H) produced by each of these bias conditions. If we plot a curve z_H as a function of z_{FA} corresponding to changes in bias, we should obtain a line with unit slope and intercept equal to d' for the fixed strength of evidence.

Now suppose that we repeat this experiment, but with a different level of evidence strength. This produces another set of pairs (z_{FA}, z_H) across the bias conditions. The second experiment is also predicted to produce a linear relation between z_H and z_{FA}, with unit slope but now with a different

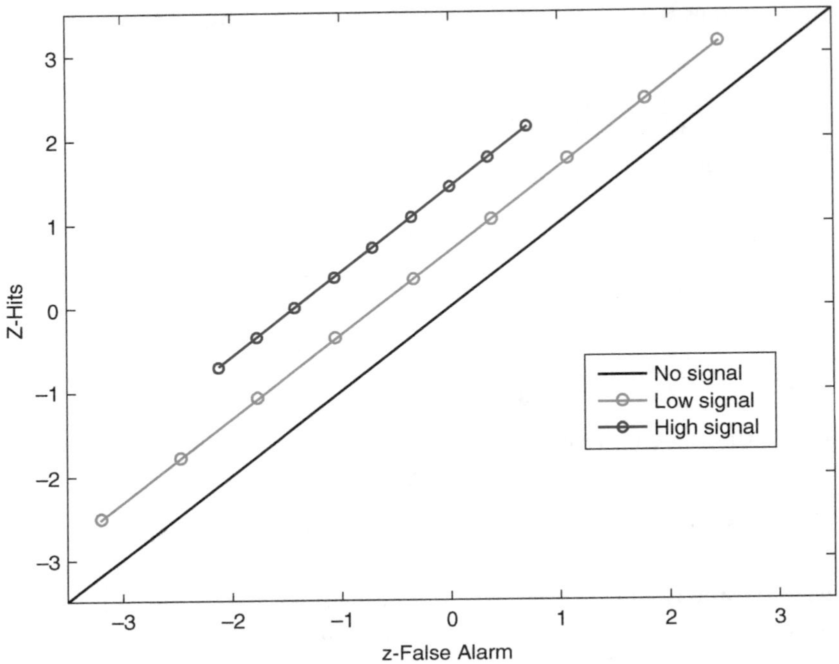

Figure 5.4
Z-ROC curve produced by plotting z_H as a function of z_{FA}, with a different curve for each signal strength.

intercept. Therefore, if we plot the lines from both experiments, with different lines for different signal strengths, then they should be parallel lines that have unit slope and intercept determined by the signal strength (see figure 5.4 for an example). For a recent example of empirical tests of Z-ROC predictions from signal detection models, see Yonelinus and Parks (2007) and Wixted (2007).

5.4 Confidence Rating Model

Confidence ratings provide additional measures of performance in a signal detection/categorization task. The usual procedure is first to ask for a choice (choose action A or B), and then rate the confidence of that choice on a rating scale (e.g., a three-point scale using L=low, M=medium, and H=high). For example, the decision-maker could decide that signal is present with high confidence (choose A with rating level H, denoted as AH). Alternatively, the decision-maker could decide that no signal is present with low confidence (choose B with rating level L, denoted as BL). This produces a table of hits and misses at three levels of confidence for the signal condition. Another similar table containing false alarms and correct rejections at each level of confidence can be obtained from the noise condition.

For the purpose of analysis, we assign the six pairs of choice and ratings (BH, BM, BL, AL, AM, and AH) six increasing likelihoods regarding the signal ($R=0 \rightarrow BH$, $R=1 \rightarrow BM$, $R=2 \rightarrow AL$, $R=3 \rightarrow AL$, $R=4 \rightarrow AM$, and $R=5 \rightarrow AH$). Tables 5.3 and 5.4 illustrate the type of results obtained with this procedure when signal is present. Another table like this is obtained when no signal is present.

The decumulative probabilities shown in table 5.4 can be computed from the joint relative frequencies in table 5.3 by summing across the appropriate cells in table 5.3. For example,

$$\Pr(R \geq 2 \mid s) = \Pr(R=2 \mid s) + \Pr(R=3 \mid s) + \Pr(R=4 \mid s) + \Pr(R=5 \mid s).$$

A corresponding table of decumulative probabilities can also be obtained for the no signal condition.

The univariate signal detection model presented here can easily be extended to choice and confidence, such as in tables 5.3 and 5.4. For this analysis, we need to associate criteria along the axis that divides the

Table 5.3

Joint relative frequency distribution for hits and misses at each confidence level when signal is present.

Signal present	Low	Medium	High
Choose A given s	$\Pr(R=3 \mid s)$	$\Pr(R=4 \mid s)$	$\Pr(R=5 \mid s)$
Choose B given s	$\Pr(R=2 \mid s)$	$\Pr(R=1 \mid s)$	$\Pr(R=0 \mid s)$

Table 5.4

Decumulative frequency distribution of hits and misses at each confidence level when signal is present.

Signal present	Low	Medium	High
Choose A given s	$\Pr(R \geq 3 \mid s)$	$\Pr(R \geq 4 \mid s)$	$\Pr(R \geq 5 \mid s)$
Choose B given s	$\Pr(R \geq 2 \mid s)$	$\Pr(R \geq 1 \mid s)$	$\Pr(\mathrm{R} \geq 0 \mid s)$

evidence into six categories. This is illustrated in figure 5.5, which shows five criteria that divide the signal distribution into the six likelihood scores (0 for *BH*, 1 for *BM*, 2 for *BL*, 3 for *AL*, 4 for *AM*, and 5 for *AH*). For example, the probability of choosing action *A* with low level of confidence corresponds to

$$\Pr(R=3 \mid s)=\Pr(R \geq 3 \mid s)-\Pr(R \geq 4 \mid s),$$

According to the model, this is the area under the curve between the criteria $c=3$ and $c=4$. This is found by first determining the area above criterion $c=3$, and then subtracting the area above $c=4$. The model states that (for $j=0, 2, \ldots, 5$),

$$\begin{aligned}\Pr(R=j \mid s) &= \Pr(R \geq j \mid s)-\Pr(R \geq j+1 \mid s) \\ &= F\left(\frac{\mu_s-c_j}{\sigma}\right)-F\left(\frac{\mu_s-c_{j+1}}{\sigma}\right),\end{aligned}$$

with $c_0=-\infty$ and $c_6=+\infty$. For example, when $j=3$, we have

$$\begin{aligned}\Pr(R=3 \mid s) &= \Pr(R \geq 3 \mid s)-\Pr(R \geq 4 \mid s) \\ &= F\left(\frac{\mu_s-c_3}{\sigma}\right)-F\left(\frac{\mu_s-c_4}{\sigma}\right).\end{aligned}$$

In this example, the bias boundaries are equally distant, but this is not a necessary assumption. The boundaries can be estimated as free parameters in the model. The entire model can be fit conveniently by using a categorical data program that is available in many statistical packages. (see https://osf.io/53ce9/ for a Bayesian estimation program in R).

5.5 A Bivariate Normal Distribution with Nonhomogenous Variance Model

Here, we will illustrate an example of a signal detection model with data that have $r=2$ dimensions. We examine the effect that nonhomogeneity has on the predictions of the model. Suppose that a radiologist examines images such as x-ray, magnetic resonance imaging (MRI), or ultrasound, and must decide if a disease is present or not. Suppose that two features are relevant: the size of a node and the brightness/darkness of a node in the image. Each feature can be represented as a random variable that takes on different values for different images. Let $X_1=$size and $X_2=$brightness corresponding to an image. When presented with an image, the decision-maker

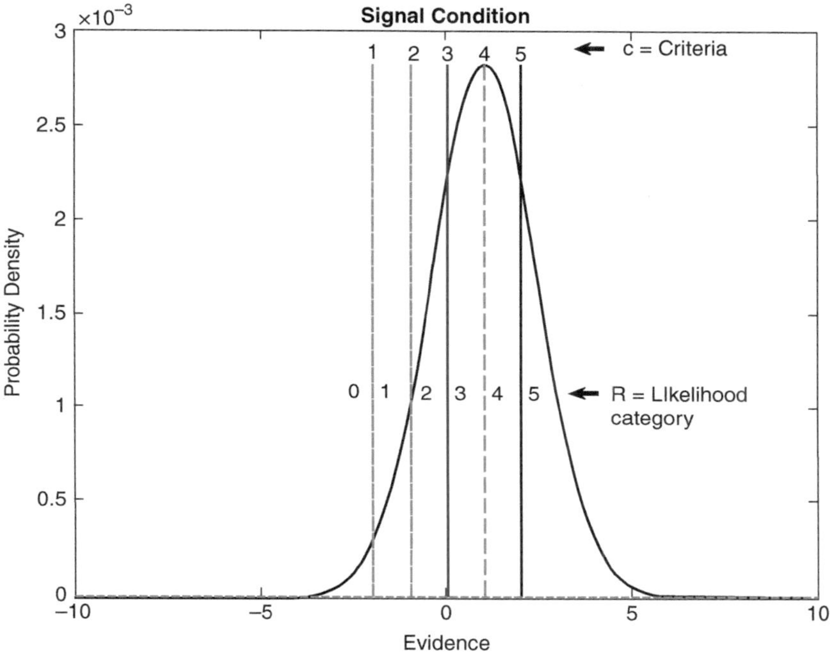

Figure 5.5
Confidence model for six likelihood scores defined by five criteria.

observes a pair of values on these two random variables and forms a unidimensional random variable that represents the evidence Y, for or against, the disease hypothesis on the basis of this pair of values. The top two panels in figure 5.6 present the contour plots of the two-dimensional distribution for the random vector $X = [X_1, X_2]$: the top panel for the signal being present, and the middle panel for no signal. The bottom panel will be discussed later in this chapter.

Each curve in the contour plot contains the pair of feature values $X = [X_1, X_2]$ that produce the same level of probability density. A small ellipse represents points with high density, and a large ellipse represents points with low density. Imagine having a bowl on top of a table, and imagine cutting into the bowl at different heights. A cut at each height produces a contour. Note that the contour plot for the signal present distribution reveals a mean located at zero, but a positive correlation between the features. The middle contour for the no signal distribution also reveals a mean

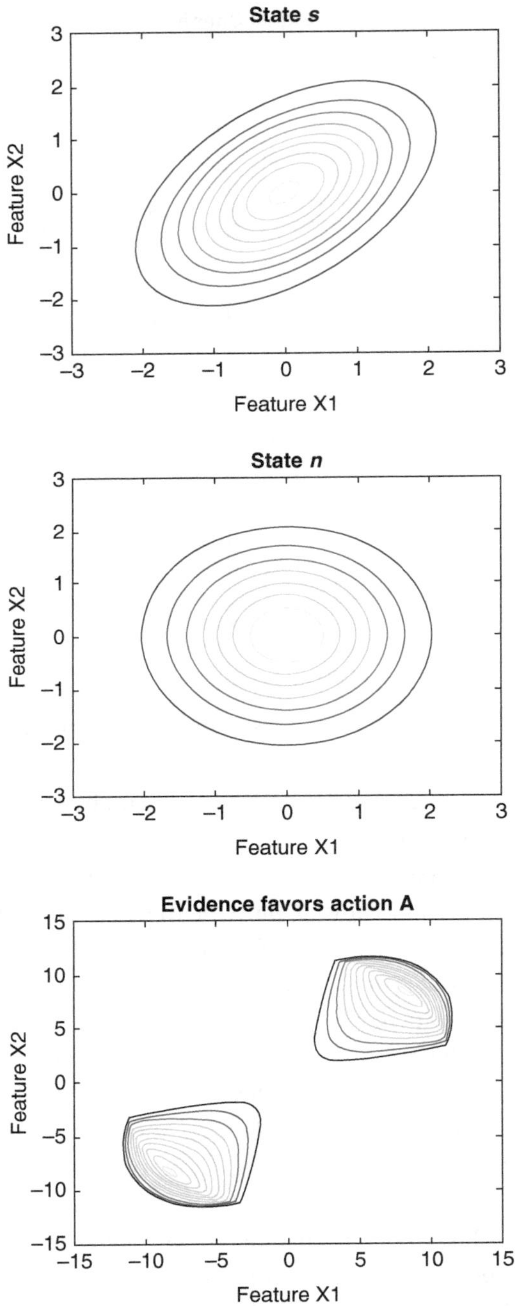

Figure 5.6
Contour plots of the signal and no signal distributions.

equal to zero, but no correlation between the features. In this case, the signal does not change the mean; instead, it induces a correlation between the features. This is an example of perceptual dependence, discussed later in this chapter (Ashby & Townsend, 1986).

In fact, both distributions were generated from a two-dimensional normal distribution with mean equal to zero and variance–covariance matrices equal to

$$\Sigma_s = \begin{bmatrix} 1 & .5 \\ .5 & 1 \end{bmatrix}, \quad \Sigma_n = \begin{bmatrix} 1 & 0 \\ 0 & 1 \end{bmatrix}.$$

The bottom panel of figure 5.6 shows the region where the log likelihood (evidence) for the signal state s is greater than that for the no signal state n. This region is located in the lobes of the upper-right and lower-left corners, where the positive correlation provides an advantage for state s over state n.

We now assume that the evidence used to make the decision, denoted Y, is based on the log likelihood ratio, so $Y = ln\ L(X)$. Although the random data X is two-dimensional in this example, the decision is based on the unidimensional random evidence Y, similar to figure 5.1. Figure 5.7 shows the distribution of this evidence for each state of nature: the top panel represents the signal present state s, and the bottom panel represents state n. The vertical line shows the median of each distribution, and the raised bar shows the 90 percent highest density interval. [This distribution was generated using the Markov chain Monte Carlo (MCMC) techniques described in chapter 4.] Even though the data vector X is bivariate normal, the evidence Y is clearly not normal at all. Also note that even though the means of the data are both zero, the median evidence under the signal state is higher than the median evidence under the no signal state.

We use the distributions of evidence shown in figure 5.7 to compute the Z-ROC curve (as done in figure 5.4 for equal variance univariate normal distributions) to obtain the graph shown in figure 5.8. The Z-ROC curve lies above the chance line because of the correlation that occurs under the signal state. This graph illustrates a striking nonlinear (positive acceleration) curvature in the Z-ROC plot. Examples of *U*-shaped Z-ROC curves such as this have appeared in the memory recognition literature (e.g., Yonelinus & Parks, 2007). In summary, this simple two-dimensional example illustrates how a bivariate normal distribution of data with nonhomogeneous covariance matrices can produce Z-ROC predictions that differ qualitatively from the pattern predicted by distributions with homogenous variance.

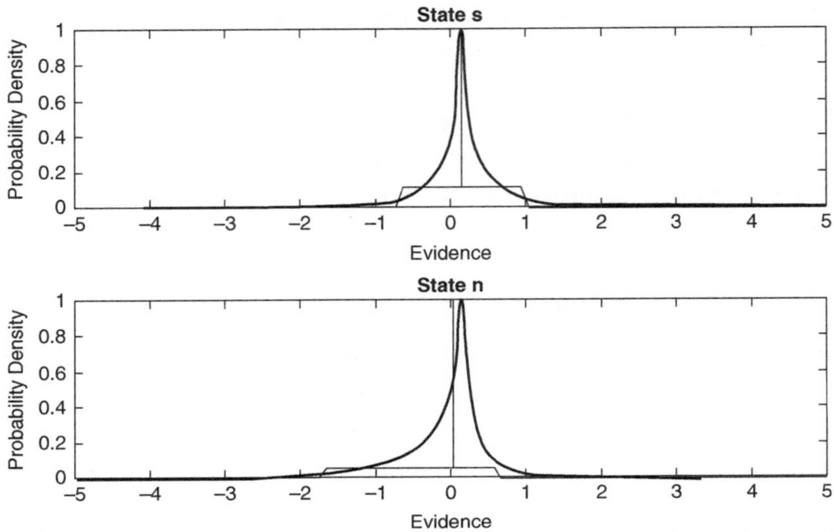

Figure 5.7
Univariate distribution of evidence produced by the joint distributions for the data shown in figure 5.6. Top panel represents state *s*, and bottom panel represents state *n*.

5.6 General Multivariate Normal Distribution Signal Detection Model

Now we assume that X is an r-dimensional, normally distributed random vector, and that

$X \sim \text{Normal}(\mu_s, \Sigma_s)$ if state s is present, and

$X \sim \text{Normal}(\mu_n, \Sigma_n)$ if state n is present,

where μ_j is the mean vector or centroid under state j and Σ_j is the variance-covariance matrix under state j. We do not necessarily assume homogeneous variance (i.e., we do not assume that $\Sigma_s = \Sigma_n$) at this point. The probability density, conditioned on state of nature i, is determined by

$$f_i(X) = (2\pi \cdot |\Sigma_i|)^{-1/2} \cdot e^{-\frac{1}{2}((X-\mu_i)^T \Sigma_i^{-1}(X-\mu_i))}.$$

The log likelihood, given a state of nature i, then equals

$$\ln f_i(X) = -\frac{1}{2} \cdot [\ln(2\pi) + \ln|\Sigma_i| + (X-\mu_i)^T \Sigma_i^{-1}(X-\mu_i)].$$

The likelihood ratio is defined as $L(X) = f_s(X)/f_n(X)$ and the log likelihood ratio equals the difference

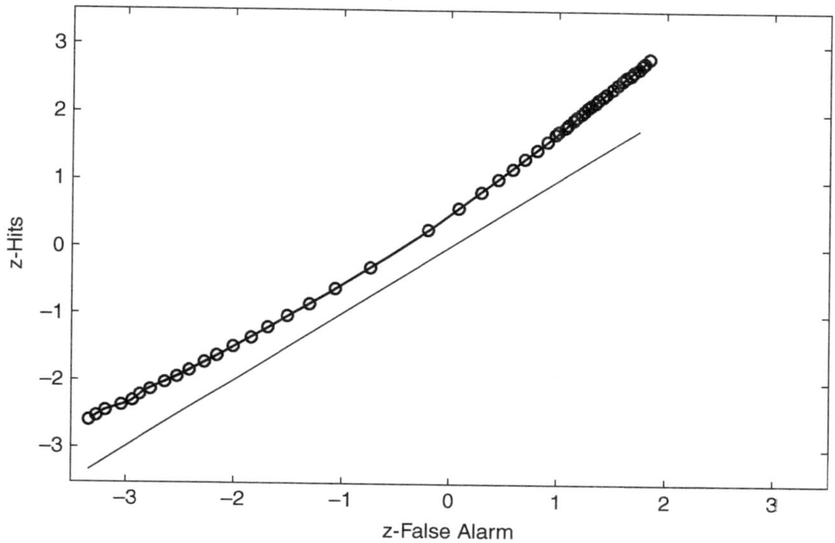

Figure 5.8
Z-ROC plot of hits and false alarms produced by the distributions of evidence shown in figure 5.7.

$$Y = \ln L(X) = \ln f_s(X) - \ln f_n(X).$$

The log likelihood ratio Y is a univariate random variable, but it is not necessarily normally distributed. In fact, the log likelihood formula given here is the same formula used to produce figures 5.7 and 5.8.

If we now assume homogenous variance–covariance matrices (assume that $\Sigma_s = \Sigma_n$), then the first two terms in the sum for the log likelihood cancel out, and the log likelihood ratio simplifies to become

$$Y = -\frac{1}{2}(X - \mu_s)^T \Sigma^{-1}(X - \mu_s) + \frac{1}{2}(X - \mu_n)^T \Sigma^{-1}(X - \mu_n)$$
$$= ((\mu_s^T - \mu_n^T)\Sigma^{-1})X + \frac{1}{2}(\mu_n^T \Sigma^{-1} \mu_n - \mu_s^T \Sigma^{-1} \mu_s).$$

This formula indicates that if $\Sigma_s = \Sigma_n$, then the log likelihood $Y = \ln L(X)$ is a linear function of the data X plus a constant. Any linear combination of normally distributed random variables is again normally distributed. Therefore, Y also is normally distributed because it is a linear combination of the normally distributed data X. Given the signal state, the mean in this case equals

$$E(Y \mid s) = (\mu_s^T - \mu_n^T)\Sigma^{-1}\mu_s + \frac{1}{2}(\mu_n^T \Sigma^{-1}\mu_n - \mu_s^T \Sigma^{-1}\mu_s)$$
$$= \frac{1}{2}(\mu_s^T - \mu_n^T)\Sigma^{-1}(\mu_s - \mu_n).$$

Likewise, the mean in the case of the no signal state equals

$$E(Y \mid n) = (\mu_s^T - \mu_n^T)\,\Sigma^{-1}\mu_n + \frac{1}{2}(\mu_n^T\,\Sigma^{-1}\mu_n - \mu_s^T\,\Sigma^{-1}\mu_s)$$
$$= -\frac{1}{2}(\mu_s^T - \mu_n^T)\Sigma^{-1}(\mu_s - \mu_n).$$

and so $E(Y \mid n) = -E(Y \mid s)$.

The variance of $Y = ln\, L(X)$ for both states equals

$$V(Y) = V(((\mu_s^T - \mu_n^T)\Sigma^{-1})X)$$
$$= (\mu_s - \mu_n)^T\Sigma^{-1}(\mu_s - \mu_n) = E(Y \mid s) - E(Y \mid n).$$

As can be seen in this formula, the variance of the evidence is twice the mean conditioned on the signal. The index of discriminability equals

$$d' = \left(\frac{E(Y \mid s) - E(Y \mid n)}{\sqrt{V(Y)}}\right) = ((\mu_s^T - \mu_n^T)\Sigma^{-1}(\mu_s - \mu_n))^{1/2}.$$

Essentially, d' equals the distance between the means in the metric of the variance–covariance matrix (i.e., when the coordinates are defined by the eigenvectors of the variance–covariance matrix). In sum, under the assumption of multivariate normal distributions with homogenous variance, the likelihood ratio distribution of signal to noise, when used to make a decision, reduces to the simple univariate model discussed in the section entitled "The Univariate Normal Homogeneous Variance Model," earlier in this chapter.

5.7 Signal Detection with Multiple (More Than Two) Categories or Actions

So far, we have restricted our attention to two actions. Now let us consider some examples with more than two actions or categories. First, we consider an example from memory recognition research (Rotello, Macmillan, & Reeder, 2004). Participants were initially presented with a list of items (e.g., words) during a training session, and then later, during a test phase, they were presented with new items that did not appear on the training list or old items taken from the training list. The participant was first asked to decide whether the test item was old or new, and if the item was chosen to be old, then the person had to say whether they remembered something specific about the item or felt as if they knew they had studied it but failed to recall any details.

One way to model this task is to assume that the participants use a single unidimensional global memory strength feature to make decisions, and that decision-makers use different cutoffs for deciding "new," "old know," and "old remember," analogous to the confidence model shown in figure 5.5.

An alternative model assumes that there are two features, as illustrated in figure 5.9: the horizontal axis of this graph represents the global item strength produced by a general but vague feeling of knowing that the item was presented, and the vertical axis represents specific item strength produced by recollection of details about the item from training. There are only two states: an old item (signal present) state *s*, which is shown as the

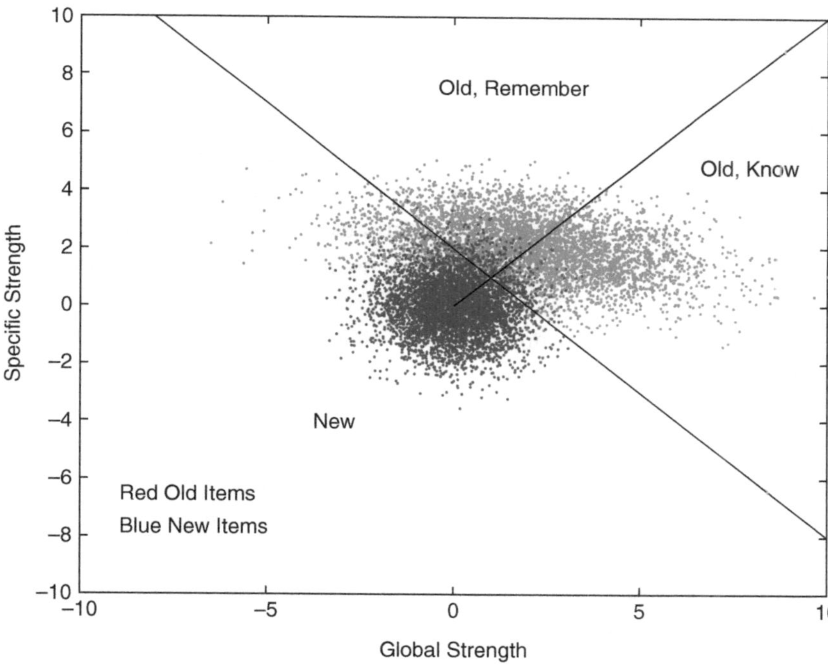

Figure 5.9
Bivariate signal detection model for the remember–know paradigm of memory recognition. Blue dots represent old item, red dots represent new item. Lines represent decision bounds. See color plate 2.

distribution of red dots, and a new item (no signal) state n, which is shown as the distribution of blue dots. However, there are two decision bounds for categorizing the items into three categories. The first bound, which is the line (specific = 2 − global) with a negative slope, represents the difference between specific and global strength. This line is used to categorize items as old versus new: items falling to the northeast of this line are categorized as old, and items falling to the southwest are categorized as new. The second bound, which is the line (specific = global) with a positive slope, represents the sum of specific and global strength. This line is used to categorize items as remembered versus known: items falling northwest of this line are categorized as remembered, and items falling southeast of this line are categorized as known. Rotello, Macmillan, and Reeder (2004) empirically tested these two competing models and found that the two-dimensional model provided a better account of the results from the remember–know memory recognition paradigm.

5.8 General Recognition Theory

One final example is a perceptual identification task (Soto et al., 2015). Suppose that there are K types of stimuli. In each trial, one type of stimulus is presented, and the person has to choose one of K categories, where each category corresponds to one of the stimulus types. This task produces a $K \times K$ confusion matrix, where each cell entry, p_{ij}, contains the probability of assigning a category in row i to a stimulus type in column j. Each stimulus type must be assigned to one of the K categories, and therefore the rows within each column of the matrix must add up to 1.

Soto et al. (2015) proposed a multivariate signal detection model called GRT-wIND for this identification task (General recognition theory with individual differences). Figure 5.10 shows an example with four stimulus types: *blue, red, green,* and *yellow*; and four categories from which to choose: Red, Blue, Green, and Yellow. In this example, the stimulus types are supposed to be designed to vary along two feature dimensions (e.g., two color mixture factors on the color wheel), so each stimulus type can be represented by a two-dimensional vector space. Using this type of design, Soto et al. (2015) then employ a bivariate normal distribution to represent each stimulus type, and they allow a different possible mean and covariance matrix for each type. Each of the four category responses is identified with

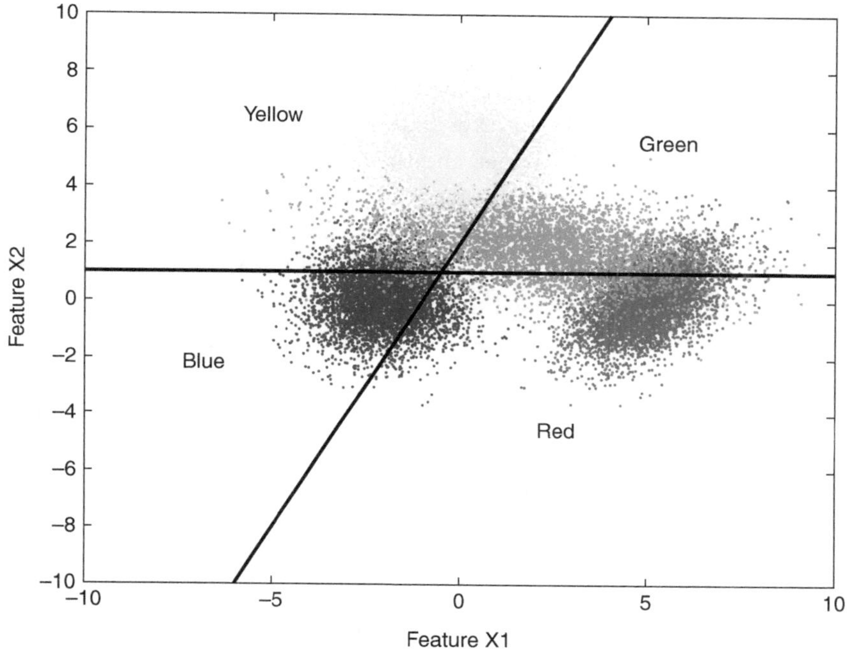

Figure 5.10
Four types of stimuli are represented by four colors. The two lines divide the space into four categories. See color plate 3.

one of the four regions formed by two linear bounds. The bounds are not necessarily optimal; instead they are restricted to linear bounds.

The GRT-wIND model assumes that all participants share a common perceptual representation (i.e., all participants use the same four distributions to represent the stimulus types). However, the model allows different individuals to employ different linear bounds. Soto et al. (2015) developed a computer program for maximum likelihood estimation of the model parameters, which is available at the website https://github.com/fsotoc/grtools. Ashby and Soto (2015) provide a more comprehensive description for how to conduct model comparisons using the GRT-wIND model. This model has been shown to produce excellent fits to confusion matrices obtained with identification tasks.

During the 1960s and 1970s, when multidimensional signal detection theory was being developed, many important concepts were introduced to describe the psychological interactions among the various features or dimensions that described a stimulus. This includes concepts of perceptual independence, decision separability, integrality, and sampling independence, to name just a few. However, these concepts were only introduced in an informal or intuitive manner, and they lacked rigorous definitions that would allow the logical relation among concepts and observable behavior to be revealed. General recognition theory (abbreviated as GRT, which is more general than GRT-wIND), was developed by Ashby and Townsend (1986), and it finally provided the rigorous theoretical framework needed to work out these important theory-to-behavior connections.

We will describe the basic ideas using a two-dimensional stimulus example, as illustrated in figure 5.10. Suppose once again that a radiologist examines images (x-ray, MRI, ultrasound, and so on), and that two features are relevant: X_1 and X_2 of a location in the image. Each perceived feature can be represented as a random variable that takes on different values for different images.

The theoretical questions concern the relations between the two features, X_1 and X_2. The two features are defined as *perceptually independent* if the two random variables are statistically independent for an image. Looking at figure 5.10, the red stimulus is perceptually dependent because there is a positive correlation between the two features; the yellow stimulus is perceptually independent because the two features are bivariate normally distributed, with zero correlation. Note that perceptual independence is symmetric: X_1 being perceptually independent of X_2 implies that X_2 is perceptually independent of X_1. Also, perceptual independence is defined for each image.

Perceptual separability refers to a different concept, which involves the manipulation of experimental factors that influence the images. For example, the images in figure 5.10 could be generated by manipulating two color mixture factors. Suppose that experimental factor B, which is supposed to influence X_2, is manipulated while keeping factor A is fixed. The manipulation of factor B, while holding factor A fixed, produces different images that vary with factor B. Factor A is said to be perceptually separable from factor B if manipulating factor B changes only the marginal distribution of feature X_2, while keeping the marginal distribution of feature X_1 constant. Suppose that in figure 5.10, yellow and blue images were formed by changing only factor B, while holding factor A constant. Clearly, the distribution along feature X_2 changes (shifts up the vertical axis). However, the manipulation of factor B also changed feature X_1 (shifting the distribution along the horizontal axis). Therefore, factor A is not perceptually independent.

Now suppose that in figure 5.10, blue and red images were formed by changing only factor A, which is supposed to influence X_1, while holding factor B constant. Clearly, the marginal distribution along feature X_1 changes (shifts right on the horizontal axis), but in this case, manipulating factor A does not change the marginal distribution of feature X_2 (the marginal distribution along the vertical axis). Therefore, based on this manipulation, factor B is perceptually independent. Note that perceptual separability is not symmetric: the effect of manipulating factor B holding A constant may not be perceptually separable, but the manipulation of factor A holding B constant may be perceptually separable. Also note that *perceptual separability* refers to a factor manipulation of the images.

A third issue concerns the boundaries that define the categories. There are two kinds of decision boundaries: separable and integral. The boundaries are decision separable if the boundaries are orthogonal to each stimulus axis (e.g., for two dimensions, the boundaries are horizontal and vertical in the figure 5.10). Otherwise, the boundaries are integral (because they require integrating information across both features X_1, X_2).

It is not possible to observe the presence or absence of these three theoretical concepts directly because they are defined with respect to the latent (subjective) features used by the decision-maker to classify images. However, the researcher can observe the probability of choosing each category conditioned on the image. What is critical for testing these theoretical properties is a task that asks the participant to categorize each feature. For example, the participant could be asked to decide whether feature X_1 is low or high, and also decide whether feature X_2 is low or high. Then it is possible to observe the joint relative frequency of responding j=low,high to the first feature and k=low,high to the second feature, denoted as $\Pr[R_1=j, R_2=k \mid \text{image}]$. Using these joint relative frequencies, we can estimate the probability of one response conditioned on another, as well as the marginal probability of one response (pooled over the other).

The implications shown in figure 5.11 have been proven for general distributions (not necessarily multivariate normal). First, we need to assume

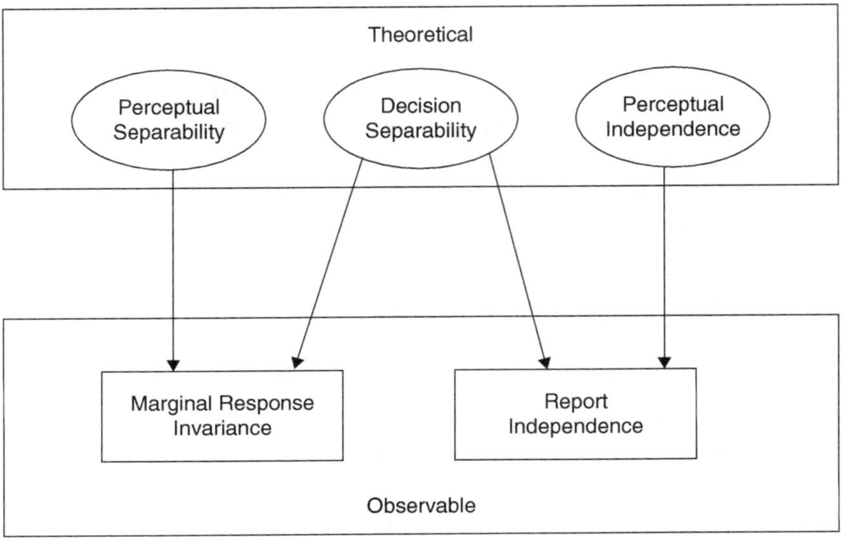

Figure 5.11
Theoretical implications of the presence of a theoretical property for the observed response relative frequencies derived from GRT.

that decision separability holds. Under the decision separability assumption, then if the responses are conditionally independent for an image, then the features are perceptually independent. Also, under the decision separability assumption, if the marginal distribution for the responses concerning feature X_1 varies only with factor A and is invariant across manipulations of factor B, then factor A is perceptually separable from factor B. The critical question concerns the assumption of decision separability. This assumption can be quantitatively tested (under parametric distribution assumptions such as multivariate normal) using the GRT-Wind program. Ever since the introduction of GRT, there have been numerous experimental tests of perceptual independence and separability (see Ashby & Soto, 2015, for a review).

5.9 Concluding Comments

The signal detection model is probably one of the most important choice theories developed in psychology. In general, it gives an account of hits and false alarm choice measures, as well as confidence ratings for signal detection types of tasks. It provides two key parameters for describing performance on a task: a sensitivity parameter *d′* and a response bias parameter *b*. It has generated widespread applications in human factors, medicine, and other applied areas. One limitation of this model, however, is its failure to take into consideration the time required to make a decision and how this decision time affects the decision outcome. This problem is addressed in the next chapter.

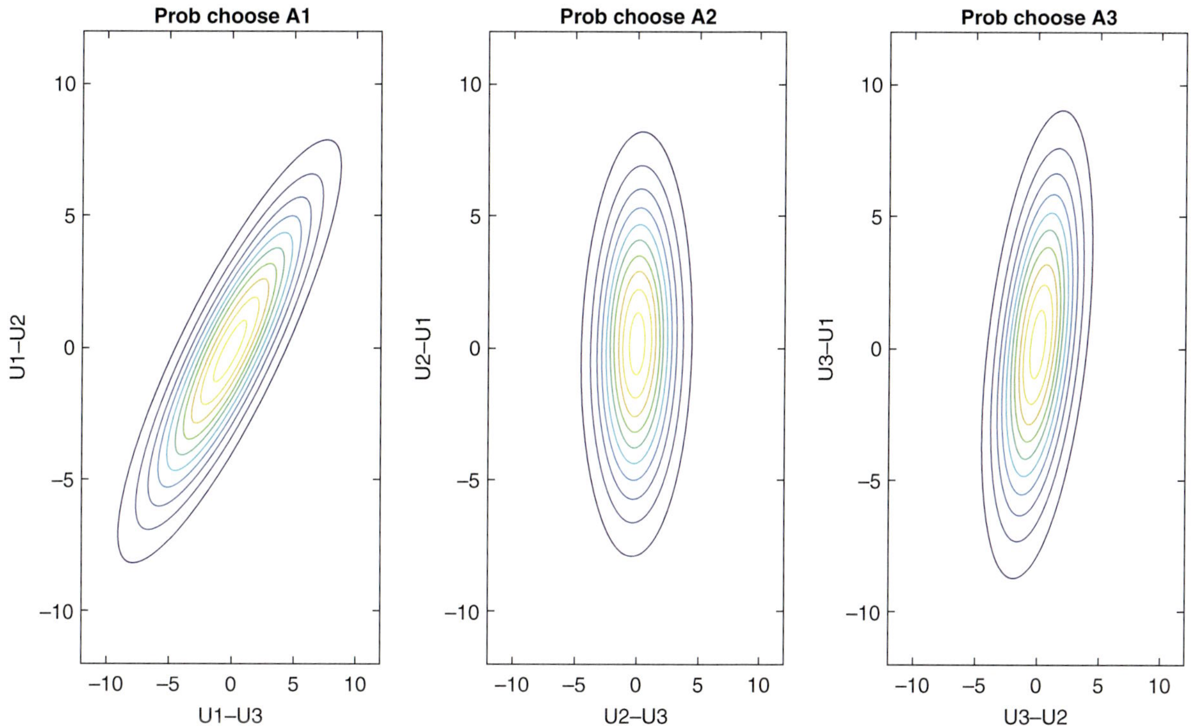

Plate 1
Bivariate joint distributions produced for each option for the triadic choice among options A1, A2, and A3.

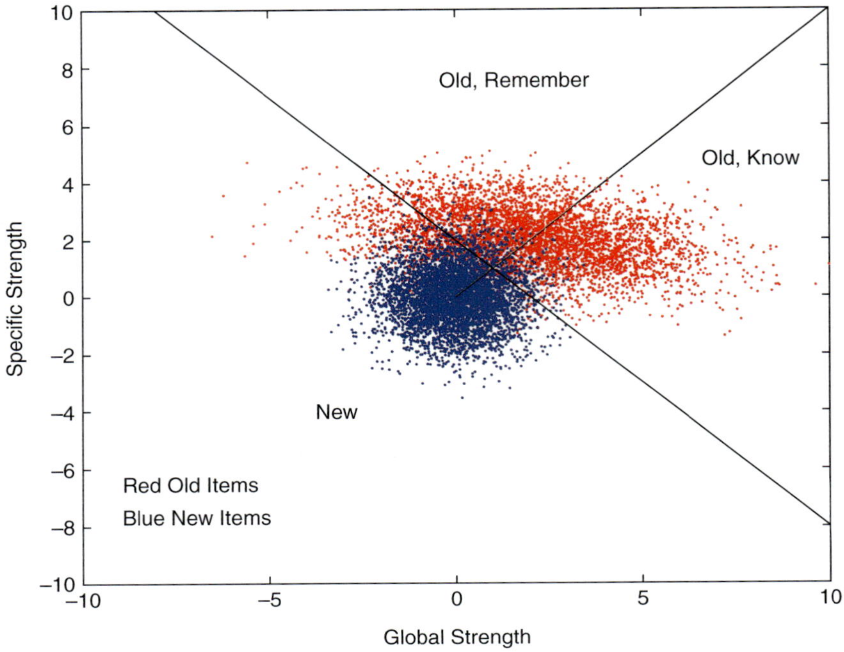

Plate 2
Bivariate signal detection model for the remember–know paradigm of memory recognition.

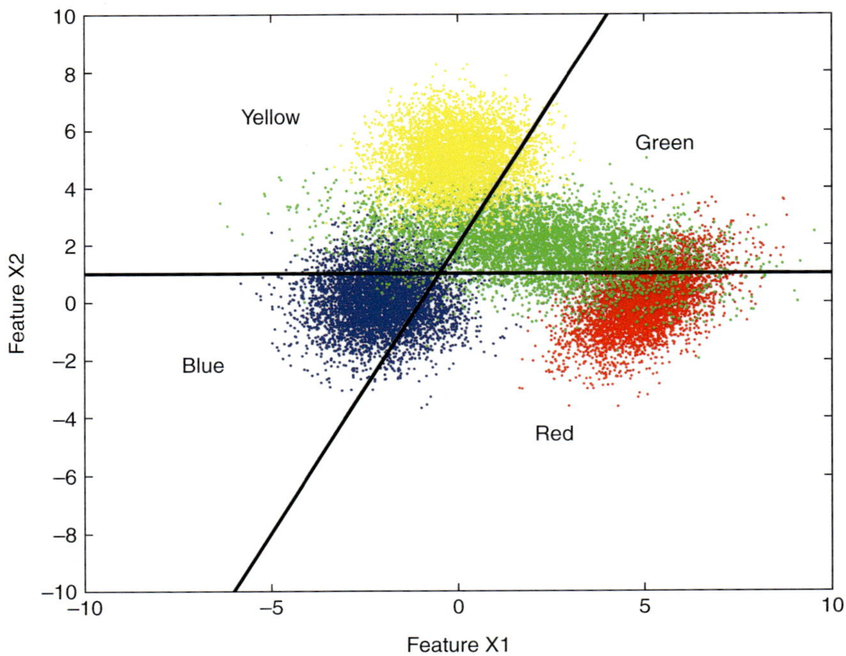

Plate 3
Four different types of stimuli. The two lines divide the space into four categories.

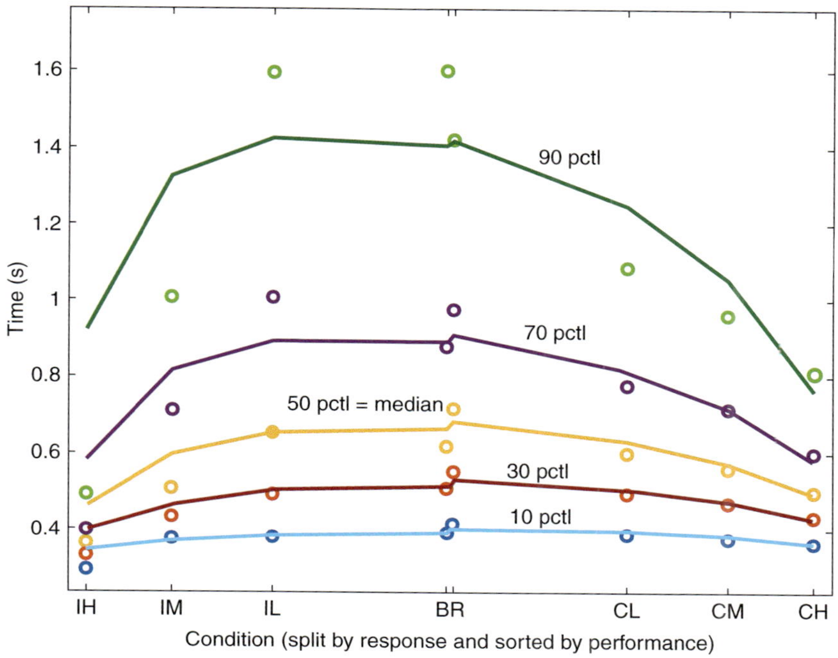

Plate 4
Quantile probability plot used to evaluate the fit of diffusion model.

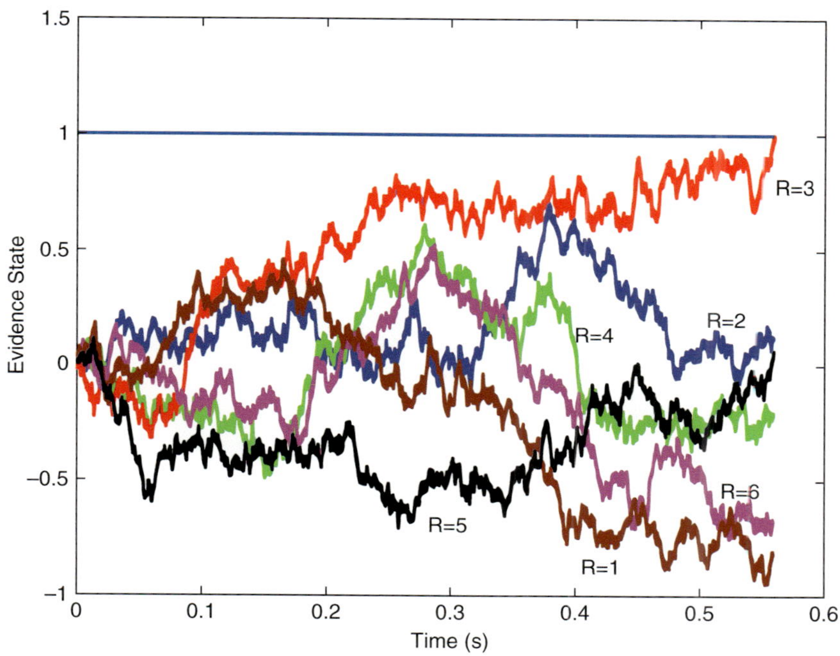

Plate 5
Separate accumulator model for directly rating confidence on a six-point confidence rating scale.

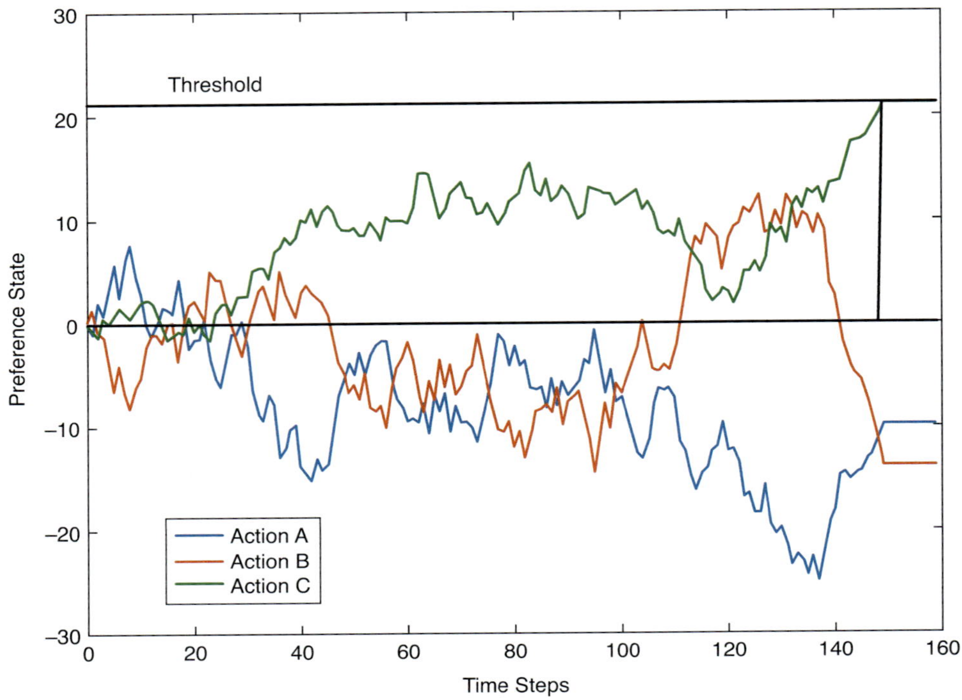

Plate 6

MDFT choice process for three alternatives. The horizontal axis represents time, the vertical axis represents preference for each action, and the flat horizontal line represents the decision threshold that has to be crossed by one of the preference states so that a choice occurs. In this case, option C first crosses the threshold at a time equal to 145 samples and is chosen by the decision-maker.

6 Sequential Sampling Models

Decisions take time and the time taken to make a decision affects the final choice. A radiologist needs time to view an image before making a diagnosis, and the time it takes to make a decision for each image is costly in view of the large number of tasks that must be performed. A military operator needs to decide whether an unidentified agent is an enemy, and must take an action before a possible attack occurs; and a juror needs time to recall and think about the prosecution and defense arguments before making a decision about the guilt or innocence of a defendant. All the models presented in the preceding chapters are static and do not take into account the time it takes to make a decision. This chapter introduces a class of models, called *sequential sampling models,* that simultaneously predict the choice and the decision time using a common psychological mechanism. The discussion primarily concerns sequential sampling models applied to signal detection and categorization types of tasks. The next chapter applies these models to preferential choice types of tasks. For more details about sequential sampling models, see Vickers (1979), Link (1992), and Luce (1986).

Sequential sampling theory was originally developed in statistics shortly after World War II (Wald & Wolfowitz, 1948). In fact, sequential analysis was used by Alan Turing to break the Enigma coding machine used by the German military during the war. Subsequently, advances were made in optimal sequential sampling theory based on Bayesian inference principles and expected utility maximization (DeGroot, 1970). Stone (1960) introduced the sequential sampling model for choice and response time to psychology. Next, Laming (1968) applied it more systematically to empirical findings in psychophysics. Link (1975) and Link and Heath (1975) generalized the model and also applied their model to perception. Vickers (1979) proposed a sequential sampling model to account for confidence in addition to choice and response time. Ratcliff (1978) developed a sequential sampling model, called a drift diffusion model, for memory recognition. Later sequential sampling models were used to account for categorization (Nosofsky & Palmeri, 1997; Ashby, 2000). Most recently, these models have been related to neural models and successfully linked to electrophysiological recordings in the brain (Bogatz et al., 2006; Gold & Shadlen, 2007; Schall, 2001). See Ratcliff et al. (2016) for a recent review of work on the diffusion model.

6.1 General Background

Suppose that a decision-maker is faced with a signal detection/categorization task like that described in chapter 5. The decision-maker must choose between two courses of action (call them action Red and action Blue), and the payoffs for an action depend on an unknown state of nature. Suppose that there are two states of nature: state s and state n (e.g., state s could represent a disease present favoring action R, and state n could represent a disease absent favoring action B). Information or data, such as an x-ray, magnetic resonance image (MRI), or ultrasound image, about the state of nature are available to the decision-maker, and using the data, the decision-maker needs to infer the state of nature in order to determine the best course of action.

The fixed sample signal detection model, described in chapter 5, assumes that a single r-dimensional data vector, X, is sampled in a batch manner (e.g., a single view of an x-ray, MRI, or ultrasound scan). Using the data X, the decision-maker infers the evidence Y regarding the state and then uses this evidence to choose the best course of action.

The sequential sampling model assumes that the decision-maker examines a sequence of data samples across time $[X_1, X_2, \ldots, X_t]$, rather than a single batch sample. For example, radiologists may not be able to comprehend a large, complex image all at once. Instead, they need to focus on specific aspects of the image, viewing a small sample from the large image at each time step. Once again, each data sample is converted to some unidimensional increment of evidence Y_t, for or against the signal state, to produce a sequence of evidence values $[Y_1, Y_2, \ldots, Y_t]$. The critical idea is that the decision-maker accumulates these evidence increments across time until the cumulative evidence becomes sufficiently strong for state s to cross a criterion to choose action R, or sufficiently strong for state n to cross a criterion to choose action *B*. When the mean evidence is positive, such that the average evidence favors action *R*, then we define action *R* as the correct action, and choosing action *B* is an error.

Note that the critical feature of the fixed sample signal detection model is that the size of the batch of features to be sampled is fixed and determined before any sample is taken. In other words, it is an exhaustive process. In contrast, the critical feature of the sequential sampling model is that the number of samples is determined in an online manner by a criterion cutoff: the process could stop early with a few samples if the early evidence is sufficiently strong, or the process could continue for a long time if the early evidence is too weak. In other words, the sequential sampling model is a self-terminating process.

The basic idea behind sequential sampling models is illustrated in figure 6.1. In this graph, the horizontal axis in the middle represents time steps (in seconds for this example) and the vertical axis represents the evidence for or against the signal state. The upper horizontal line represents the criterion amount of evidence, denoted as $+S^*$, that is required to stop and decide to choose action *R* (in favor of state *s*), and the lower blue horizontal line represents the criterion amount of evidence, denoted as $-S^*$, required to stop and decide to choose action *B* (in favor of the state *n*). The jagged trajectory represents a sample path of cumulative evidence across time. Each point on the trajectory, which we denote as $S(t)$, represents the cumulative evidence at time step *t*. In this example, the evidence hit the criterion for choosing *R* first. The point at which the sample path touches the upper criterion determines the stopping time (indicated by the vertical line projecting down from the sample path to the middle time line at $t=0.35$ seconds). In this example, the action *R* is chosen at time $t=0.35$ seconds. In sum, the probability of choosing the Red action is determined by the probability that the cumulative evidence $S(t)$ hits the top bound $+S^*$ before the bottom bound $-S^*$, and the distribution of decision time is determined by the first passage time distribution to hit a bound.

Figure 6.2 shows twenty simulations (black trajectories) of the same process. Most of the black sample trajectories hit the top bound (choose *R*), but some of them hit the bottom bound (choose *B*). This is because the threshold is rather small in magnitude, which permits errors to occur. Many of the trajectories take a short time to hit the bound, but some of them take a much longer time. The linearly increasing straight line on the graph shows

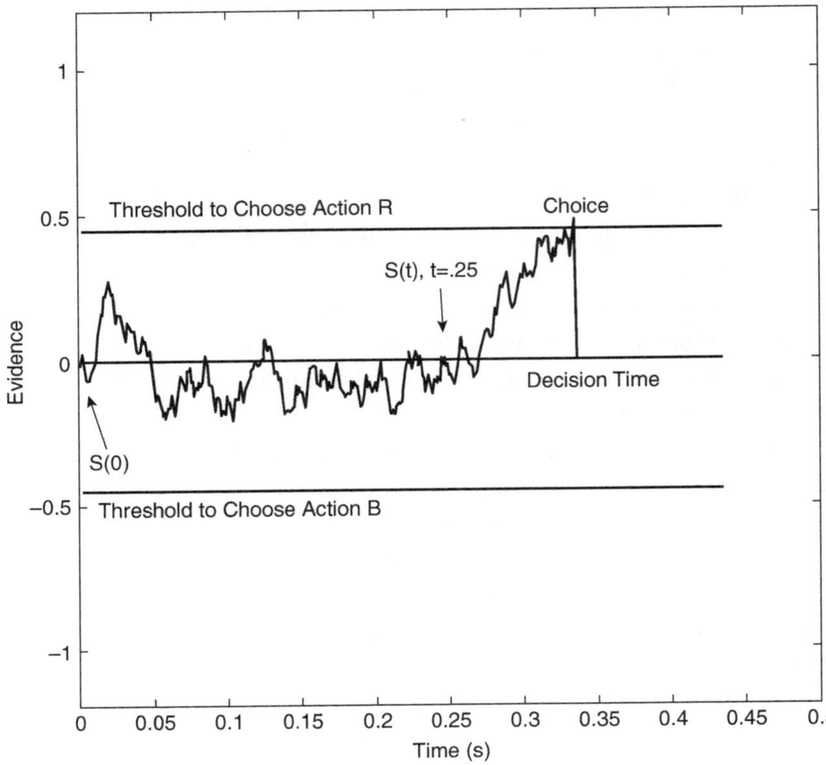

Figure 6.1
Basic idea of the sequential sampling choice process. Horizontal axis is time and vertical axis is evidence. Trajectory shows evidence accumulated across time. Top line is bound for choosing *R*, and bottom line is bound for choosing *B*.

the mean trajectory that would be obtained by averaging across a very large number of simulations. The slope of this mean trajectory is called the *mean drift rate*. The variability in the black trajectories around this mean is produced by what is called *diffusion rate*.

Figure 6.3 shows the effect of increasing the threshold bounds. As can be seen, the chance of making an error is much reduced. However, the time required to reach a decision is much longer.

The sequential sampling model has four key parameters. The first parameter is the initial starting position, *S*(0) (the prior evidence at time zero before sampling begins). Like the signal detection model, this is a bias parameter, which is influenced by prior probabilities and payoffs.

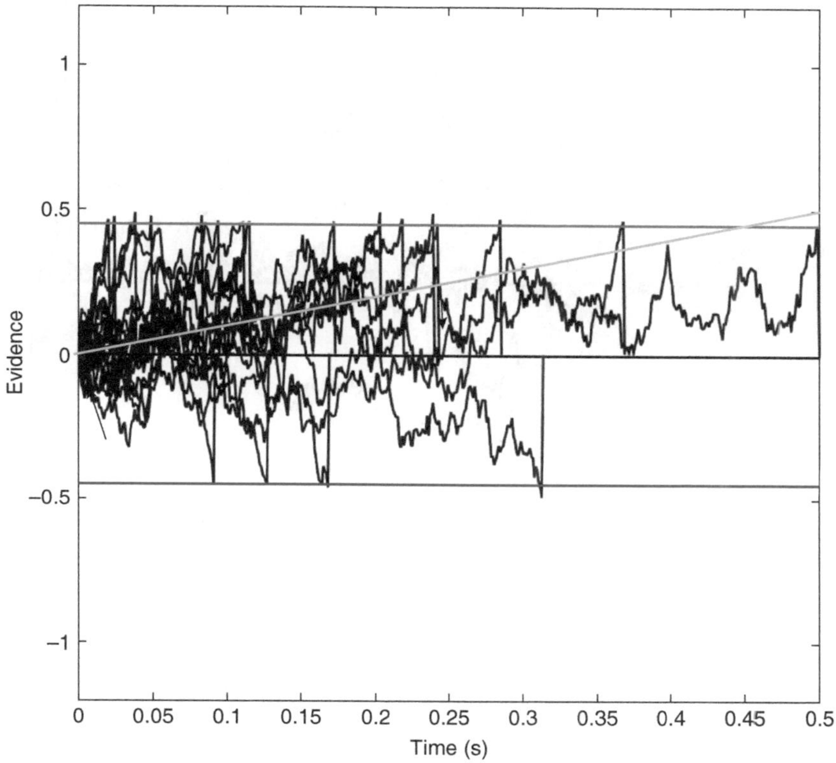

Figure 6.2
Twenty sample trajectories of the same process shown in figure 6.1.

The second parameter is the mean drift rate, which determines the mean evidence on each time step. In figures 6.1, 6.2, and 6.3, the mean evidence is positive favoring state *s*, leading to more frequent choices of action *R* (the correct action in this case).

The third parameter is the diffusion rate, which determines the variability of the evidence at each time step. Increasing the diffusion rate increases the noise and the likelihood of making an error (choosing action *B* in this case).

Finally, the amount of evidence, S^*, required to stop and make a decision is the fourth parameter. This criterion is also called the *threshold*, and it controls what is called the *speed/accuracy trade-off*. If the threshold is set to a low magnitude, then the accumulation generally hits the threshold very quickly with a small number of samples, but the small sample

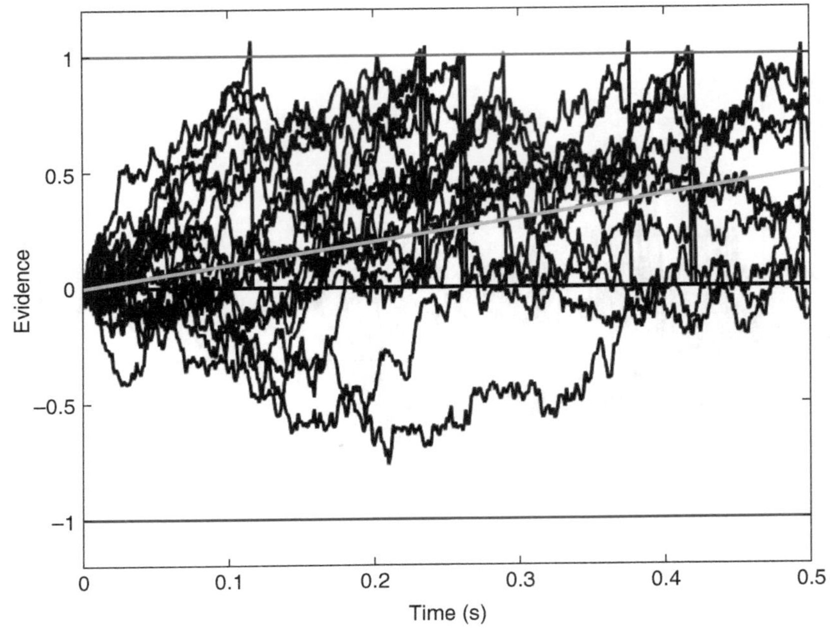

Figure 6.3
Effect of increasing the threshold on the sequential sampling process.

size will generate many errors. If the threshold is set to a high magnitude, then the accumulation generally takes a large number of samples to hit the threshold, but the large sample size will produce greater accuracy. The threshold is affected by time pressure, the cost of sampling information, and penalties for errors. Decision-makers who are very deliberative will set a high threshold, and decision-makers who are very impulsive will set a low threshold.

One might question the use of criterion bounds that are equidistant from zero. However, this places no real constraint on the model because any unequidistant bound model can be transformed into an equidistant bound model by changing the starting state S(0). Only the distance between the starting point and each bound matters.

6.2 Stochastic Difference Equation Model

Let us consider how to simulate a sequential sampling process (cf. Higham, 2001). Define h as a time step (e.g., 1 millisecond). Suppose that $X(t)$ is the data vector sampled at time t. It is assumed that this data vector was sampled under the presence of either state s or state n. Based on knowledge of the likelihood of the data given each state, data $X(t)$ is converted into an increment of evidence $Y(t)$, where positive increments represent evidence favoring state s and negative increments represents evidence favoring state n.

We will assume that $Y(t)$ is an independent and identically distributed sequence of evidence increments. Usually, but not necessarily (see Link, 1992), it is assumed that the increments are normally distributed. The mean of the evidence increment sampled at each time point depends on the state of nature that is generating the sequence of data vectors. Define $\delta_s \cdot h = \mathrm{E}(Y(t) \mid s)$ as the mean conditioned on state s for a time step h, and $\delta_n \cdot h = \mathrm{E}(Y(t) \mid n)$ as the mean conditioned on state n for a time step h. The dependence of the mean on the time step represents the idea that the mean increases proportional to the time step. The evidence increment at each time point also has a variance: $\mathrm{V}(Y \mid s) = h \cdot \sigma_s^2$ conditioned on state s, and $\mathrm{V}(Y \mid n) = h \cdot \sigma_n^2$ conditioned on state n. The variance is also assumed to increase proportionally with time step h. The evidence increment at each time point can be broken into two parts: $Y(t) = \delta_i \cdot h + \varepsilon(t) \cdot \sqrt{h}$, where the residual increment $\varepsilon(t)$ is assumed to have a mean equal to zero and variance $V(\varepsilon(t)) = \sigma_i^2$.

Define $S(0)$ as the initial bias of the accumulator. A neutral or unbiased accumulator has an initial value equal to zero, $S(0) = 0$; positive values represent an initial bias favoring action R and negative values represent an initial bias favoring action B. The decision-maker accumulates the evidence over time as follows:

$$S(t+h) = S(t) + Y(t+h) = S(t) + \delta_i \cdot h + \varepsilon(t) \cdot \sqrt{h}.$$

Finally, the stopping rule can be stated as follows:

If $S(t) \geq +S^*$, then stop and choose action R.

If $S(t) \leq -S^*$, then stop and choose action B.

Otherwise, continue sampling.

This is the rule used to generate the sample paths shown in figures 6.1 and 6.2. The simulation was generated with $h = .001$, $S(0) = 0$, $\delta = 1$, $S^* = .45$, and $\varepsilon(t)$ ~Normal(0, $\sigma^2 = 1$).

For the next point, it will be useful to rearrange the evidence accumulation into the difference equation form. If we define the change in accumulation as

$$dS(t+h) = S(t+h) - S(t),$$

then we arrive at the stochastic difference equation

$$dS(t) = \delta_i \cdot h + \varepsilon(t) \cdot \sqrt{h}.$$

When h is a finite interval, we call the accumulation process a *discrete time random walk process*. Note that the random walk is a discrete time and continuous state Markov process because the new state depends only on the previous state and a new independent increment. The trajectory produced by the random walk will be polygonal, but it becomes smoother looking as the time interval h approaches zero.

If we allow h to approach zero in the limit, then this random walk process converges on a continuous time and continuous state Markov process, which is called a *diffusion process*. A diffusion process has continuous trajectories in the following sense, called the *Dynkin condition*:

$$\frac{\Pr(|S(t+h) - S(t)| > \varepsilon \mid S(t) = x)}{h} \to 0.$$

as $h \to 0$, such that displacements greater than $\varepsilon > 0$ are very unlikely in very small time intervals.

Researchers in the field of cognitive science and neuroscience call the stochastic difference equation model that we presented the *drift-diffusion model*. In the stochastic processing literature, this model is called the *Wiener process*. The Wiener process is only one very special type of diffusion process; there are many others, such as the Ornstein–Uhlenbeck (OU) process (see chapter 7). But the Wiener model is the simplest and currently the most popular diffusion model for cognitive scientists. Ratcliff (1978) introduced cognitive scientists to the drift-diffusion model. For a recent review of applications to cognitive science, see Ratcliff et al. (2016).

We could use this simulation method to determine the choice probabilities and distribution of stopping times for the diffusion model. However, we would have to run the simulation thousands of times to obtain estimates of the choice probabilities and decision time distributions. Instead, we can theoretically derive these quantities for the Wiener version of the diffusion model.

6.3 Choice Probability and Distribution of Stopping Times for the Wiener Diffusion Model

Based on the diffusion process described in the previous section, using tools from stochastic processing theory (e.g., Feller, 1968, chapter 15;

Bhattacharyia & Waymire, 1990, chapter 5), we can mathematically derive the probability that the cumulative evidence hits the top bound for choosing the Red action before hitting the bottom bound for choosing the Blue action. We can also derive the first passage time distribution to a bound. The appendix of this chapter describes how these derivations are made; but for now, we will simply provide the solutions. We will continue with the example illustrated in figure 6.1 in which the decision-maker has a choice between two actions, R and B.

Recall that σ_i^2 is the diffusion rate for the process for condition i. For the remainder, let us focus on one condition (say $i=s$), and drop the subscript i. To simplify the formulas, it is convenient to standardize the other three parameters with respect to the diffusion rate as follows:

$d=(\delta/\sigma)$ standardized mean drift rate

$\theta=(S^*/\sigma)$ standardized threshold bound

$b=S(0)/\sigma$ standardized starting bias

The parameter d is analogous to the d' used in signal detection theory, except that here it refers to the rate of evidence accumulation. The parameter b is analogous to the bias parameter in signal detection theory. However, the threshold parameter θ is unique to the sequential sampling model.

Using stochastic processing theory, we can derive the probability of choosing action R, which is equal to

$$p(R)=\frac{e^{4\cdot d\cdot\theta}-e^{2\cdot d\cdot(\theta-b)}}{e^{4\cdot d\cdot\theta}-1}. \tag{1}$$

Note that if $b=0$, then the choice probability reduces to the well-known logistic model

$$p(R)=\frac{e^{4\cdot d\cdot\theta}-e^{2\cdot d\cdot\theta}}{e^{4\cdot d\cdot\theta}-1}=\frac{e^{2\cdot d\cdot\theta}(e^{2\cdot d\cdot\theta}-1)}{(e^{2\cdot d\cdot\theta}-1)(e^{2\cdot d\cdot\theta}+1)}=\frac{e^{2\cdot d\cdot\theta}}{(e^{2\cdot d\cdot\theta}+1)}=\frac{1}{(1+e^{-2\cdot d\cdot\theta})}.$$

The probability density to choose action R at time t can be derived and shown to be equal to

$$p(t,R)=\frac{\pi}{(2\theta)^2}\cdot e^{d\cdot(\theta-b)}\cdot\sum_{j=1}^{\infty}j\cdot e^{-\frac{1}{2}\left(\left(\frac{\pi\cdot j}{2\theta}\right)^2+d^2\right)\cdot t}\cdot\sin\left(\frac{\pi\cdot(\theta-b)\cdot j}{2\theta}\right). \tag{2}$$

Note that $p(R)=\int_0^\infty p(t,R)\,dt$, so the distribution sums to the choice probability for action R. The probability density to choose action B at time t can be obtained from the same formula after changing the sign of the standardized mean drift rate from d to $-d$, changing the sign of the initial starting point from b to $-b$.

Figure 6.4a shows examples of predictions from the Weiner diffusion model using equations (1) and (2). The predictions shown in figure 6.4a were derived using $\theta = 1$ and $b = 0$ and two levels of discriminability ($d = .50$ and $d = .75$). The low drift rate produced a probability of choosing action R equal to .73, and the high drift rate produced a higher probability, equal to .82. As can be seen in figure 6.4a, the distributions are generally skewed to the right, and variance and the skew increases with the mean of the response time. Decreasing the mean drift rate shifts the median and mode to slower decision times.

Figure 6.4b shows the effect of increasing the threshold from $\theta = 1$ to $\theta = 1.5$ when $d = .5$ and $b = 0$. Even though we see the same change in accuracy (probability of choosing Red), the change in distribution produced by changing the threshold is quite distinct from the change in the mean drift rate. Increasing the threshold not only increases the median, but also increases the initial rise in the distribution and lengthens the tail of the distribution.

Finally, figure 6.4c shows the effect of changing the bias from zero to $b = .2$ in favor of action R, with $d = .5$ and $\theta = 1$. As can be seen, increasing the bias in favor of action R increases the probability of choosing that action, and decreases the median time to choose that action. Also, the leading edge for the rise in the distribution is shorter under the bias.

It is interesting to examine what would happen if the data were generated from the Weiner diffusion model, but we analyzed the data from the point of view of the static signal detection model. In particular, consider what happens to the discriminability and bias when we change the threshold of the diffusion model. Note that the threshold is a cognitive control parameter determining decision caution, and it should not reflect discriminability. Table 6.1 shows the results of this analysis.

The first three columns of the table show the parameters used to generate the predictions of the diffusion model. The fourth column shows the probability of choosing R that is produced by these parameters. The fifth column shows the corresponding z-hits and z–false alarms produced by taking the inverse normal transform. The sixth column shows the signal detection estimate of $d' = (z_H - z_{FA})$ and $b = (z_H + z_{FA})/2$.

As can be seen in table 6.1, simply changing the threshold substantially increased the signal detection estimate of the discriminability. However,

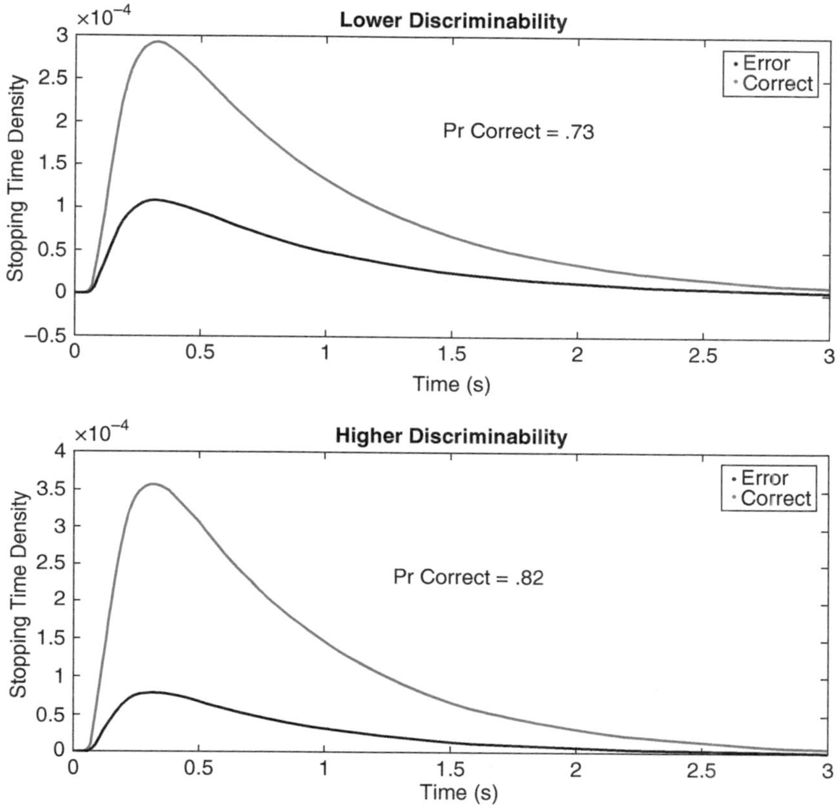

Figure 6.4a

Effect of changing the mean drift rate on the distribution of decision time for each action ($d = .50$ for lower drift rate, $d = .75$ for higher drift rate).

according to the diffusion model, discriminability is indexed by the drift rate parameter d (strength of evidence per unit time step), which did not change. The threshold, which did change, reflects a change in decision caution (i.e., taking time to collect more evidence) rather than discriminability (i.e., strength of evidence per unit time).

6.4 Nondecision Time Component

One other factor that needs to be taken into consideration is *nondecision time,* an extra amount of time required for initial encoding of the decision

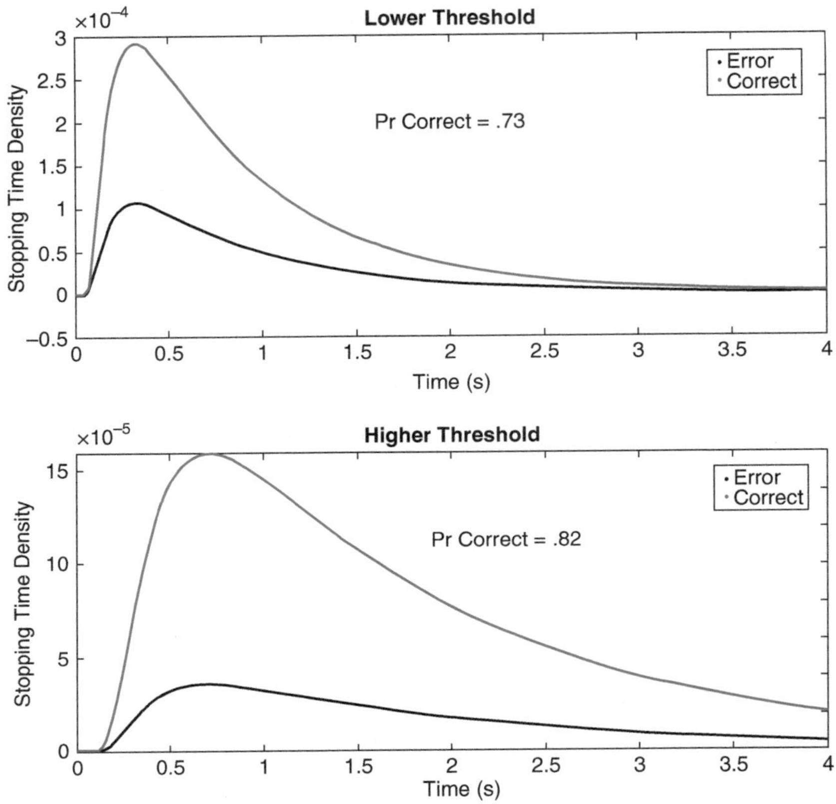

Figure 6.4b
Effect of changing the threshold on the distribution of decision time for each action ($\theta = 1$ for lower threshold, $\theta = 1.5$ for higher threshold).

problem and for making motor response. This time is assumed to be unrelated to the time required for evidence accumulation. This nondecision time is often assumed to be uniformly distributed, which is a convenient but not empirically plausible assumption. Ratcliff (2013) has argued that the model fit is fairly robust with respect to these distribution assumptions when nondecision time is small compared to decision time. Verdonck and Tuerlinckx (2016) have developed tools for deconvoluting nondecision time from decision time to obtain an approximately uncontaminated decision time from the raw decision time.

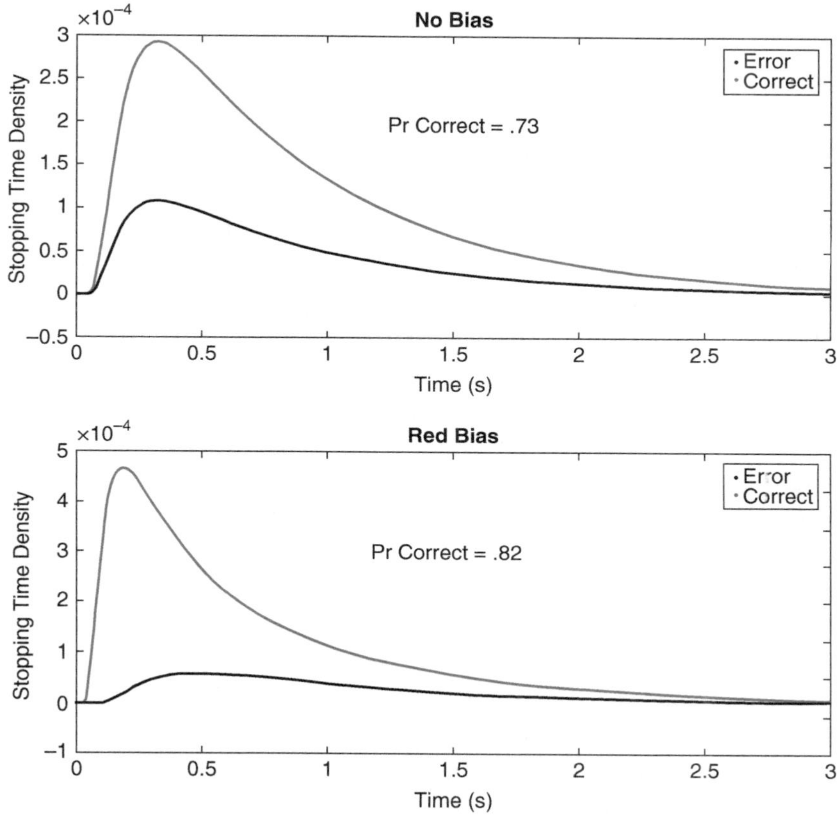

Figure 6.4c
Effect of changing the bias on the distribution of decision time for each action ($b=0$ for no bias, $b=.2$ for red bias).

Table 6.1
Changes in d' and bias for static signal detection model produced by changes in the threshold θ generated by the diffusion model.

Diffusion Model Parameters					
d	b	θ	$p(R)$	Signal Detection Analysis	
.50	.20	1.0	0.8082	$z_H=.871$	$d'=1.22$
−.50	.20	1.0	0.3631	$z_{FA}=-.350$	$b=.26$
.50	.20	2.0	0.9058	$z_H=1.315$	$d'=2.35$
−.50	.20	2.0	0.1497	$z_{FA}=-1.04$	$b=.14$

6.5 Parameter Variability Assumptions

The Weiner diffusion model has one serious problem. Recall that if the mean drift rate is positive, so the average evidence favors action *R*, then we define action *R* as the correct action, and choosing action *B* is an error. If the bias equals zero, then the model predicts that the distribution of stopping time conditioned on choosing action *R* equals the distribution of stopping time conditioned on choosing *B*. These two distributions are predicted to be identical, even though the choice probability is much higher for action *R* than action *B* (Ratcliff, 2013). Empirically, these distributions are not equal, and thus some changes are required in the Weiner model. With speed stress and easy tasks, errors tend to be faster than correct responses; under accuracy stress and difficult tasks, errors tend to be slower than correct responses (Vickers, 1979; Luce, 1986).

One way is to change the assumptions regarding the accumulation process (Link, 1975; Vickers, 1979). For example, slow errors occur if the accumulation process is represented by a race between two imperfectly correlated accumulators (Vickers, 1979, Usher & McClelland, 2001). Another way is to change the assumption that the threshold bounds remain constant across time, and instead allow the bounds to decrease across time (Shadlen & Kiani, 2013). Decreasing threshold bounds also tends to produce slower errors. A third way is to assume that the mean evidence changes across time from strong to weak (Diederich & Oswald, 2016). A fourth way is to assume that the parameters of the Weiner process vary across trials (Ratcliff, 1978).

Ratcliff (1978) argues that the mean drift rate parameter varies across trials, and this has become one of the most common ways to account for differences in choice response time distributions for correct and error trials. This parameter variability might be caused by uncontrolled variations in the stimulus from trial to trial, or by uncontrolled variation in attention by the decision-maker. Ratcliff (1978) assumes that the mean drift rate is normally distributed across trials. Variability in the mean drift rate tends to produce slower errors. To produce fast errors, Ratcliff (2013) allows the starting point to vary across trials; it is usually assumed to be uniformly distributed. Thus, different patterns of slow and fast errors are produced by balancing the effect of starting point and mean drift rate variability.

There are now numerous programs available on various websites for estimating the parameters of the drift diffusion model with parameter variability.

The general idea is to fit the parameters to the joint density of the choice and time taken to make a decision based on equations (1) and (2). Some of the programs provide maximum likelihood estimation (e.g., Vandekerckhove & Tuerlinckx, 2008); others provide Bayesian estimation (e.g., Weicki, Sofer, & Frank, 2013, Vandekerckhove, Tuerlinckx, & Lee, 2011). Ratcliff and Childers (2015) provide a comprehensive review and evaluation of these various programs.

6.6 Methods for Empirically Testing the Model

On the one hand, the drift diffusion model, which includes the variations in parameters across trials, has been criticized as being too flexible, and possibly untestable (Jones & Dzhafarov, 2014). This occurs when one is allowed to arbitrarily choose distributions over parameters. On the other hand, this same model, using the distribution assumptions described previously, has undergone some of the most rigorous quantitative empirical tests ever conducted in cognitive science, and it has proved to be impressively successful (Ratcliff & Smith, 2015). Next, we describe how to test the drift diffusion model empirically. To illustrate this method, we generated simulated data from the drift diffusion model using the Vandekerckhove and Tuerlinckx (2008) program and using the following hypothetical experiment.

Suppose that participants are asked to rapidly decide whether a large random display of dots came from a population that had a majority of red or blue dots. This experiment includes 200 trials per condition for each participant, and half the trials contain a sample with a majority of red dots, and the remaining half contain a majority of blue dots. The proportion of dots is manipulated across four conditions to vary the strength of evidence (the mean drift rate) across four conditions: 50 percent, 55 percent, 60 percent, or 65 percent majority dots. The model parameters are then estimated separately for each participant from the 1,000 response time and choices made on the 250 trials apiece of the four conditions.

A very effective way to test and evaluate the fit of the diffusion model empirically is to examine what is called a *quantile probability plot*. Figure 6.5 shows an example of the hypothetical experiment. This type of graph contains a lot of information in a compressed manner. First, consider the horizontal axis. Each tick mark represents a combination of a response (C for correct versus I for incorrect) and a condition (zero, low, medium, or high

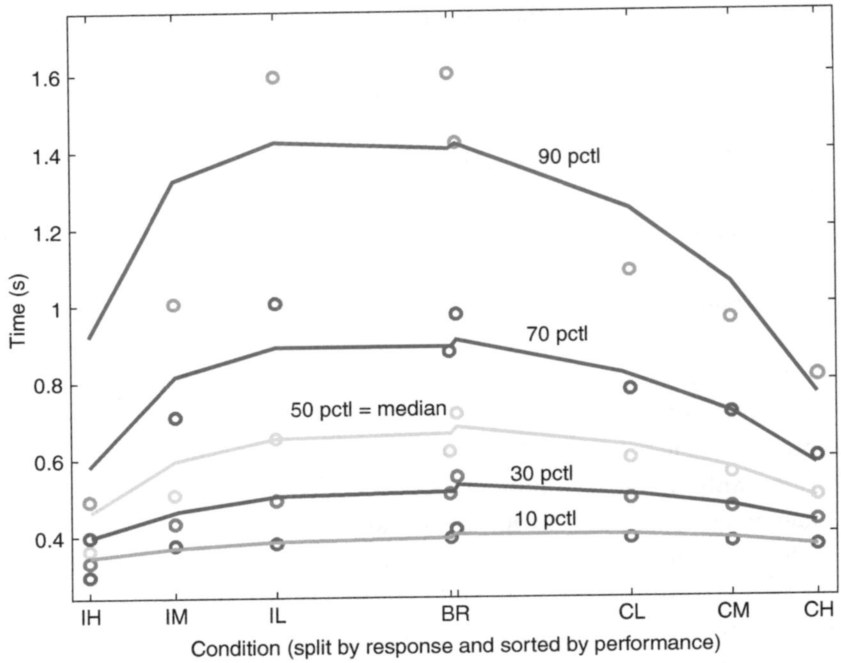

Figure 6.5
Quantile probability plot used to evaluate the fit of a diffusion model. See plate 4.

evidence condition). For example, the tick mark labeled IL indicates the incorrect response (either a *B* response when *R* was the majority, or an *R* response when *B* was the majority) for the low evidence condition (55 percent majority). This tick mark is located at the proportion of trials that this incorrect response occurred at this condition (approximately, 0.25).

Likewise, the tick mark labeled CL indicates the correct response (either a *B* response when *B* was the majority, or an *R* response when *R* was the majority) for the low evidence condition (55 percent majority). This tick mark is located at the proportion of trials that this correct response occurred at this condition (approximately, 0.75). The vertical axis shows the quantiles of the conditional response time distribution. Each curve plots a quantile as a function of the response-condition combination plotted on the horizontal axis. For example, the yellow middle dots and curve show the observations and predictions for the median response time for each response-condition combination. The other curves show the 10th, 30th, 70th, and 90th percentiles.

For example, the lowest blue dots and curve show the 10th percentile, which is the leading edge of the response time distribution. Likewise, the highest green dots and curve show the 90th percentile, which is the tail of the response time distribution.

As can be seen in this graph, the model does a fairly good job of tracking the quantile probability points. This is expected because the data were simulated from the model. The deviations are produced by sampling with small sample sizes. However, fits similar to this are frequently reported in applications of the drift diffusion model to provide convincing support for the model (see, e.g., Ratcliff et al. 2016 for a review).

6.7 Confidence Judgments

The drift diffusion model has been extended to account for confidence in addition to choice and response time. Consider a task in which the participant is presented a stimulus and first makes a binary choice (signal present or absent), and then rates the confidence of this choice on a 0–100 probability scale (0 means sure that the signal is not present, and 100 means sure that signal is present). Postdecision confidence models have been developed for this task (Van Zandt & Maldonado-Molina, 2004; Pleskac & Busemeyer, 2010; Moran, Teodorescu, & Usher, 2015), one of which is illustrated in figure 6.6. Essentially the same ideas described earlier are used to model the initial binary choice and choice response time. To model the subsequent confidence rating, it is assumed that the decision-maker continues to process the information (either from the display or from memory of the display) for some postdecision period of time. The final position of evidence in postdecision processing is used to determine the final confidence rating.

Next consider a task in which the participants are presented with a stimulus and directly provide a confidence rating regarding a hypothesis (e.g., how confident they are that the presented item is an old item from a memory list). Figure 6.7 illustrates this idea for a six-point confidence rating scale (ranging from 1 = sure it is not old, to 6 = sure that it is old). Ratcliff and Starns (2009, 2013) proposed a model for this type of task that assumes that each confidence rating value has a separate evidence accumulator. The accumulator that first reaches its decision boundary determines which confidence rating is chosen.

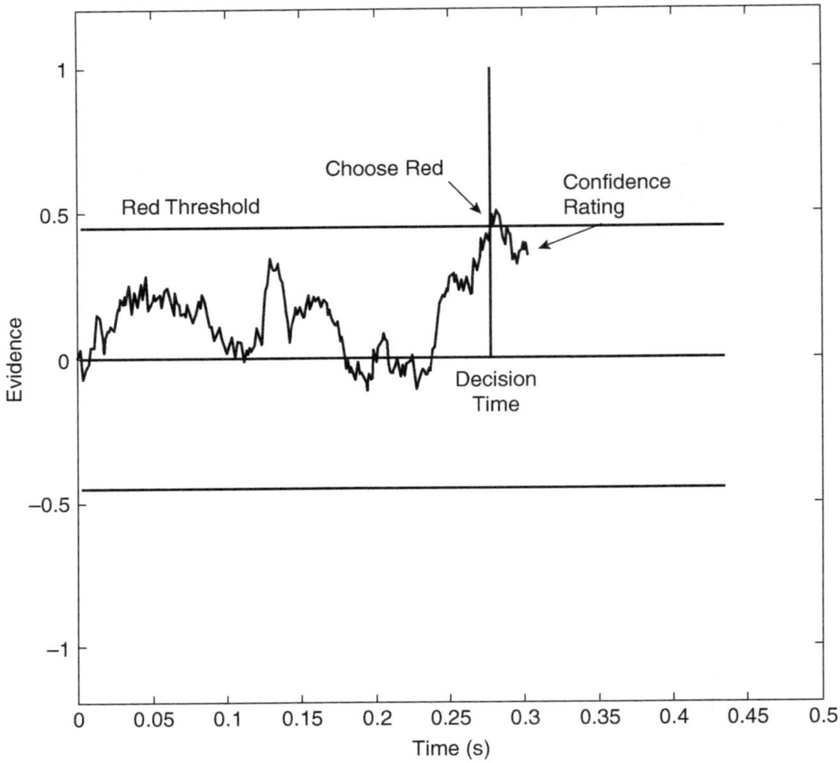

Figure 6.6
Model of binary choice, response time, and later confidence rating. The confidence rating corresponds to the final point on the evidence scale that occurs after postdecision processing indicated by the arrow.

In their confidence model, an increase in the evidence in one accumulator is accompanied by a decrease in the others, with the total amount of evidence in the system being constant. This allows their confidence model to reduce to the binary choice diffusion model when there are only two possible confidence levels (e.g., decide if an item is old or new). When there are only two levels, the two accumulators become perfectly negatively correlated (e.g., the accumulator for "new" equals the negative of the accumulator for "old"). If one accumulator (e.g., the one for "old") reaches a negative threshold level (indicating to reject old), then this implies that the other accumulator (e.g., the one for "new") has reached the positive bound (to accept as new). Therefore, only one accumulator (e.g., the "old"

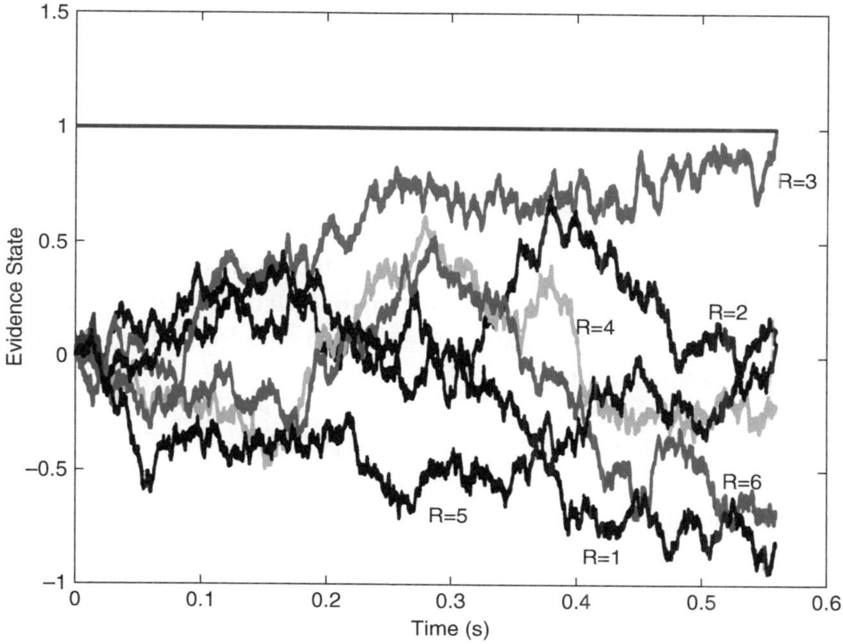

Figure 6.7
Separate accumulator model for directly rating confidence on a six-point scale. See color plate 5.

accumulator) is needed to determine the final choice and response time, and figure 6.7 (with only two accumulators) can be reduced to figure 6.1.

6.8 Alternative Accumulation Models

The Ratcliff diffusion model for binary choice is probably one of the most popular evidence accumulation models, but there are many other competitors. Next, we briefly describe two alternative ways to model evidence accumulation.

LCA model Usher and McClelland (2001) developed a model for evidence accumulation, called the *Leaky Competing Accumulator (LCA)*. This model is similar to the Ratcliff confidence model (figure 6.7) because it assigns a separate accumulator to each alternative. The accumulators can be negatively correlated, but even in the binary choice case, they are not perfectly negatively correlated. Therefore, even in the binary choice case, two separate accumulators race to determine the choice. This makes the model more complex. Even binary choices must be simulated in a two-dimensional plane. No closed-form solutions exist for the LCA model. However, computational methods for fitting the model have been developed (Turner & Van Zandt, 2018).

Define $S_i(0)$ as the initial state of evidence for alternative i, and $S_i(t)$ is the accumulated state evidence for alternative i. These states accumulate across time, according to a stochastic accumulation process described here, until a threshold S^* is reached and the first option to reach the threshold is chosen.

The LCA model accumulates evidence according to the following stochastic difference equation:

$$S_i(t+h) = S_i(t) - \left(\gamma \cdot S_i(t) + I_i - \sum_{k \neq i} \beta \cdot S_k(t)\right) \cdot h + d\varepsilon(t+h),$$

where $d\varepsilon(t)$ is a Brownian motion noise term, $\beta > 0$ is a constant lateral inhibition, γ is a leakage of evidence factor, and I_i is the input mean evidence. Unlike the Ratcliff diffusion model, the LCA model truncates the evidence state equal to zero if the next computed step on the right side of this equation happens to fall below zero any time point t.

One advantage of the LCA model is that it can predict differences in mean choice time for correct and incorrect choices without requiring any mean drift rate variability assumptions. Second, the lateral inhibitory factor allows the LCA model to produce recency (i.e., leakage of evidence over time). Usher and McClelland (2001) provide some evidence for these kinds of recency effects.

LBA model Another well-known model is the Linear Ballistic Accumulation (LBA) model by Brown and Heathcote (2008). This model is very different from the stochastic accumulation models. According to this model, each action corresponds to an accumulator, they race in parallel, and the accumulator that first reaches threshold S^* is chosen. However, once a starting position ($S_i(0)$, which is the intercept of a line at time zero), and the "drift rate" (d_i, which is equal to the slope of a line) are selected for each action, the accumulators deterministically integrate the slope over time $S_i(t) = S_i(0) + d_i \cdot t$. Thus, the time to reach the threshold for each action is simply the distance over rate: $T_i = (S^* - S_i(0))/d_i$. The action with this shortest time is chosen. At this point, the whole model is deterministic. To make the model probabilistic, it is assumed that the starting point and drift rate vary across the trial according to some specified distributions. The deterministic accumulation of evidence may not be a realistic psychological or neurophysiological assumption, but it does greatly simplify the model calculations. Also, the LBA model fits the data about as well as Ratcliff's diffusion model, and the parameters provide similar interpretations.

6.9 Continuous Response Models

Most recently, extensions of diffusion models to continuous response tasks have been developed (Ratcliff, 2018; Smith, 2016; Kvam, 2019). Figure 6.8 illustrates one idea for a continuous percentage rating scale. For example, suppose that a participant is asked to judge the likelihood that a presented item is an old item from a memory list by moving a mouse to a location along a semicircular arc on a screen with percentages ranging continuously from 0–100. The mouse trajectory may look like figure 6.8. To be clear, the continuous response models do *not* assume that a mouse trajectory directly reflects the decision process. However, some of the continuous response models *do* hypothesize that an internal (e.g., neural) stochastic process, like the stochastic process used to determine figure 6.8, is used to obtain the observed final value reported on the continuous response scale. At each moment in time, the internal state $S(t)$ is represented by a position located in the plane representing the state space. The evidence provided by the stimulus, plus noise, stochastically drive the state through the plane until the state reaches a specific threshold bound. In figure 6.8, the state drifted up and hit the semicircular bound around the location corresponding to a percentage judgment equal to 60 on the arc.

6.10 Concluding Comments

Sequential sampling models are becoming the most dominant and important decision models in cognitive science. Like the signal detection model, sequential sampling models provide a coherent account of choice and confidence. But beyond the signal detection model, sequential sampling models also account for decision time and the effect of time pressure on choice. Like the signal detection model, sequential sampling models provide measures of discriminability and bias to account for choice. But beyond signal detection models, sequential sampling models also provide a measure

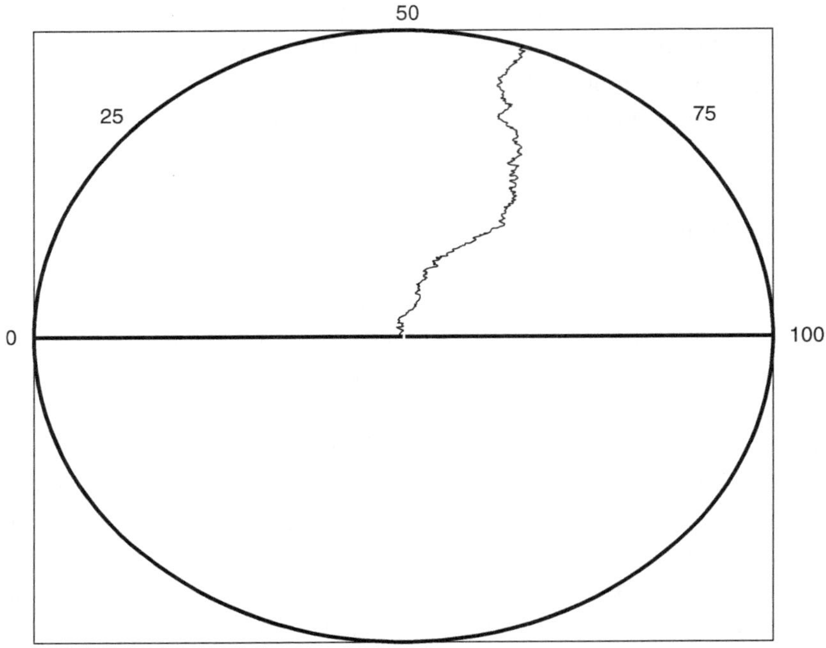

Figure 6.8
Continuous response model for judging the probability of an event on a continuous scale ranging from 0–100.

of response caution, represented by the threshold bound, when making a choice. The threshold bound is critical for understanding the speed and accuracy trade-off in decision-making.

There are many competing versions of these models, but this chapter has focused on the Wiener diffusion model (i.e., the drift diffusion model), which is the most popular implementation of sequential sampling models. Now that computer software is available for fitting the drift diffusion model, it is beginning to have wider applications outside of cognitive science to clinical fields, developmental psychology, and aging research. This discussion has been solely concerned with signal detection/categorization type applications in which the sequential sampling process accumulates evidence regarding some unknown state before taking an action. The next chapter describes how sequential sampling models can be applied to preferential choice tasks in which the sequential sampling process accumulates affective evaluations in order to make decisions.

6.A Appendix

This appendix provides a derivation of choice and response time predictions for two alternative drift diffusion Weiner model.

For a completely rigorous derivation of the prediction equations [equations (1) and (2) presented earlier in this chapter] for the Wiener process, see Bhattacharya & Waymire (1990, chapter 5). However, we can outline the main steps in order to provide some insight into the derivations. Assume that $S(t)$ is a Wiener process, and define $p_t(x, y)$ as the probability density of transiting from value $S(0)=x$ to value $S(t)=y$ during time interval t, while staying inside the boundaries $-\theta<y<+\theta$. Let $f(x)$ be a smooth function of x; in particular, $f(x)$ could be interpreted as $p_t(x, y)$ with variable x for fixed values of y and t.

The derivation starts with the definition of the conditional expectation $E(f(S(t))\,|\,S(0)=x)$ of a future value of $f(S(t))$ given a previous known start value of $S(0)=x$. Now the tricky part is to apply a Taylor series expansion to function $f(S(h))$ for a very small h, which produces a very small variation of $S(h)$ around the known starting value $S(0)=x$:

$$f(S(h)) \approx f(x) + (S(h) - x)\frac{df(x)}{dx} + \frac{1}{2}(S(h) - x)^2 \frac{d^2 f(x)}{dx^2}.$$

Only the second order is needed because any higher-order terms are assumed to go to zero in the limit according to the definition of a diffusion process (see Bhattacharya & Waymire, 1990, p. 368, definition 1.1).

Taking expectations of this expansion, and substituting the mean $E(S(h)-x\,|\,S(0)=x)=h\cdot\delta$ and the variance $E((S(h)-x)^2\,|\,S(0)=x)=h\cdot\sigma^2$ of the diffusion process into this expectation, produce

$$E(f(S(h))\,|\,S(0)=x) \approx f(x) + h\cdot\delta\frac{df(x)}{dx} + \frac{1}{2}h\cdot\sigma^2\frac{d^2 f(x)}{dx^2}.$$

Finally, we take the limit

$$\lim_{h\to 0}\left(\frac{E(f(S(h))\,|\,S(0)=x) - f(x)}{h}\right) = \delta\frac{df(x)}{dx} + \frac{1}{2}\sigma^2\frac{d^2 f(x)}{dx^2}.$$

This limit is called the *Kolmogorov backward equation*, which is then used to solve for the prediction equations.

First, consider equation (1), which gives the probability of choosing action R over B, given that the process started at position x, denoted here as $p(R\,|\,x)$. This can be derived from the backward equation as follows. Define

$P_h(R \mid x)$ as the probability of starting from x and taking a step to $-\theta<y<+\theta$ within a small time h, and then eventually choosing action R from y. Using the law of total probability, this can be expressed in terms of the transition density $p_t(x, y)$ from x to y as

$$P_h(R \mid x) = \int p_h(x, y) \cdot P(R \mid y)dy = E(P(R \mid S(h)) \mid S(0) = x).$$

This equation is a special case of the conditional expectation operator with $f(S(t))=P(R \mid S(t))$. Therefore, $P_h(R \mid x)$ satisfies the Kolmogorov backward equation. In this case, $\lim_{h\to 0} \frac{P_h(R \mid x) - P_0(R \mid x)}{h} = 0$ [see Karlin & Taylor, 1981, p. 193, equation (3.1)], we obtain the equality

$$0 = \delta \frac{\partial P(R \mid x)}{\partial x} + \frac{1}{2}\sigma^2 \frac{\partial^2 P(R \mid x)}{\partial x^2}.$$

If we solve this homogeneous, second-order linear differential equation for $P(R \mid x)$, along with the boundary conditions, $P(R \mid +\theta)=1$ and $P(R \mid -\theta)=0$, then we obtain equation (1).

Next, we need to determine equation (2). To do that, we need to show that $p_t(x, y)$ satisfies the backward equation. According to the Chapman–Kolmogorov equation for Markov processes,

$$p_{t+h}(x, y) = \int p_t(z, y) \cdot p_h(x, z)dz = E\big(p_t(S(h), y) \mid S(0) = x\big).$$

So, defining $f(S(h))=p_t(S(h), y)$ and inserting this into the limit produces

$$\lim_{h\to 0}\left(\frac{p_{t+h}(x, y) - p_t(x, y)}{h}\right) = \frac{\partial p_t(x, y)}{\partial t} = \delta \frac{\partial p_t(x, y)}{\partial x} + \frac{1}{2}\sigma^2 \frac{\partial^2 p_t(x, y)}{\partial x^2}.$$

This last result is the Kolmogorov backward equation for the probability density (which is termed *backward* because it is based on a function of the starting value $S(0)=x$).

The problem now becomes solving the backward equation for $p_t(x, y)$. This can be done using spectral decomposition methods (Bhattacharya & Waymire, 1990, chapter 5). For a fixed y and t, the probability density $p_t(x, y)$ is a function of x, which is an element of a Hilbert space of density functions. We can define A as a linear operator that maps one function p in the Hilbert space onto another function q, such that $Ap=q$. In particular, the Kolmogorov backward equation is a linear mapping defined by

$$Ap = \left[\delta \frac{\partial}{\partial x} + \frac{1}{2}\sigma^2 \frac{\partial^2}{\partial x^2}\right] \cdot p$$

$$Ap(x) = \delta \frac{\partial p(x)}{\partial x} + \frac{1}{2}\sigma^2 \frac{\partial^2 p(x)}{\partial x^2}.$$

The Hilbert space includes an inner product, $\langle p \mid q \rangle$. For this application, it is useful to define the inner product as

$$\langle p \mid q \rangle = \int p(x) \cdot q(x) \cdot \omega(x)\, dx$$

for elements p, q of the Hilbert space, where the weighting function for the inner product is defined as $\omega(x) = e^{2d\left(\frac{x}{\sigma}\right)}$. This inner product was chosen because, with respect to this product, the linear operator $A = \left[\delta \frac{\partial}{\partial x} + \frac{1}{2}\sigma^2 \frac{\partial^2}{\partial x^2}\right]$ is symmetric. Because it is symmetric, it can be described by a spectral representation in terms of eigenvectors and eigenvalues. Define ψ as an eigenfunction of A and λ as the corresponding eigenvalue, so that $A\psi = \lambda\psi$. Then $(e^{\lambda t}\psi)$ is a solution to the backward equation:

$$\frac{\partial}{\partial t}(e^{\lambda t}\psi) = e^{\lambda t}(\lambda\psi) = e^{\lambda t} \cdot A\psi = A \cdot (e^{\lambda t}\psi).$$

Every eigenvector of A provides a solution, and furthermore, any linear combination of solutions of A forms another solution:

$$\begin{aligned} \frac{\partial}{\partial t} \sum_{j=0}^{\infty} \langle \psi_j | p \rangle \cdot (e^{\lambda_j \cdot t}\psi_j) &= \sum_{j=0}^{\infty} \langle \psi_j | p \rangle \cdot (A \cdot e^{\lambda_j \cdot t}\psi_j) \\ &= A \cdot \sum_{j=0}^{\infty} \langle \psi_j | p \rangle \cdot (e^{\lambda_j \cdot t}\psi_j). \end{aligned}$$

So now we need to determine the eigenvectors and eigenvalues of the linear operator A.

There are many solutions to the backward equation, and picking a unique solution requires defining the boundary conditions for a process with two absorbing bounds. In the case of absorbing bounds, the backward equation must satisfy the Dirichlet boundary conditions:

$$\lim_{x \downarrow -\theta} p_t(x, y) = 0, \;\; \lim_{x \uparrow +\theta} p_t(x, y) = 0.$$

The solutions for the eigenvectors that satisfy the eigen-equation $A\psi = \lambda\psi$, as well as the boundary conditions, are

$$\psi_j(x) = \frac{1}{\sqrt{\theta \cdot \sigma}} e^{-d \cdot \left(\frac{x}{\sigma}\right)} \sin\left(\frac{j \cdot \pi}{2\theta} \cdot \frac{x}{\sigma}\right),$$

and the corresponding eigenvalues are

$$\lambda_j = -\frac{1}{2}\left(\left(\frac{j \cdot \pi}{2\theta}\right)^2 + d^2\right).$$

The eigenvectors provide a basis for spanning the Hilbert space. A function in this space can be represented as a linear combination of the eigenvectors.

Define q_s as the transition density function with values $p_s(y, z)$ describing the transition from y to z during time interval s. Consider the following solution for the transition probability density from x to z during time $t+s$:

$$p_{t+s} = \sum_{j=0}^{\infty} e^{\lambda_j \cdot t} \cdot \langle \psi_j | q_s \rangle \cdot \psi_j.$$

More specifically, this transition density at a value of x equals

$$p_{t+s}(x, z) = \sum_{j=0}^{\infty} \left(\int p_s(y,z) \cdot \psi_j(y) \cdot \omega(y) \cdot dy \right) \cdot e^{\lambda_j \cdot t} \psi_j(x)$$

$$= \int p_s(y, z) \cdot \left(\sum_{j=0}^{\infty} \psi_j(y) \cdot e^{\lambda_j \cdot t} \cdot \psi_j(x) \cdot \omega(y) \right) \cdot dy.$$

This solution satisfies the Kolmogorov backward equation (because it is a linear combination of solutions to the equation), and it also satisfies the Chapman-Kolmogorov initial conditions by equating

$$p_t(x, y) = \sum_{j=0}^{\infty} \psi_j(y) \cdot e^{\lambda_j \cdot t} \cdot \psi_j(x) \cdot \omega(y).$$

Inserting the eigenvectors and eigenvalues into this final solution for the transition density, then, equals

$$p_t(x, y) = \sum_{j=1}^{\infty} \psi_j(x) \cdot e^{\lambda_j t} \cdot \psi_j(y) \cdot \omega(y)$$

$$= \frac{1}{\theta \cdot \sigma} e^{d \cdot \left(\frac{x-y}{\sigma} \right)} \sum_{j=1}^{\infty} e^{-\frac{1}{2}\left[\left(\frac{j \cdot \pi}{2\theta} \right)^2 + d^2 \right] \cdot t} \sin\left(\frac{j \cdot \pi}{2\theta} \cdot \frac{x}{\sigma} \right) \cdot \sin\left(\frac{j \cdot \pi}{2\theta} \cdot \frac{y}{\sigma} \right).$$

Recall that $p_t(x, y)$ is the transition density for the process while staying inside the boundaries $-\theta < y < +\theta$. Define $P(R \mid x)$ as the probability that R is chosen over B, given that the process started at x. We now wish to derive the cumulative probability, denoted as $P(t,R \mid x)$, of stopping at or before time t and choosing action R. This equals the probability of eventually choosing action R minus the probability of reaching the point y without stopping before time t, but then eventually choosing action R from point y. Therefore, we obtain

$$P(t, R \mid x) = P(R \mid x) - \int_{-\theta}^{\theta} P(R \mid y) \cdot p_t(x, y) dy.$$

The derivative of this cumulative probability provides the probability density to stop at time t and choose R, given the start at position x:

$$p(t, R \mid x) = -\int_{-\theta}^{\theta} P(R \mid y) \cdot \frac{\partial}{\partial t} p_t(x, y)\, dy.$$

Inserting the derivative of $p_t(x, y)$ into this expression and then integrating finally produces (after much algebraic manipulation) the probability density equation described earlier in this chapter [equation (2)].

7 Decision Field Theory

Preferential choices, such as consumer choices, medical decisions, and military decisions, are often made by taking time to think about the advantages and disadvantages of each course of action. The models in this and the next chapter attempt to describe this deliberation process using the following shared ideas: A decision-maker needs to consider various attributes, features, or possible states of the word, which takes time to consider and evaluate. The competing options are compared according to their value for each attribute or feature. These comparisons produce changes in evaluation for each action that are accumulated across time into a preference state for each action. This preference state evolves across time, increasing in preference for options that are advantageous on an attribute under consideration and decreasing in preference for other options that are disadvantageous at that moment. The accumulation of preference continues until the preference for one of the options becomes sufficiently strong that a threshold is crossed and a decision is made.

Chapter 3 presented random utility models. This and the next chapter can be viewed as dynamic extensions of those models. Rather than sampling a single random utility and making a choice, as assumed in chapter 3, we now assume that a sequence of random utilities is sampled and accumulated so that the decision-maker can make a choice.[1] The number of samples now depends on how quickly one of the options reaches threshold. The dynamics that describe the sequential sampling and accumulation of preference are very similar to those described in chapter 6. The main difference concerns the nature of the accumulation: in chapter 6, evidence for or against each hypothesis was accumulated over time; in this chapter and the next chapter, affective evaluations for or against competing courses of action are accumulated over time. However, the formal (mathematical) properties of the models are very similar.

The earliest work on dynamic models of preference was by Marley and Colonius (1992), who proposed independent "horse race" models for making preferential choices. Later, Busemeyer and Townsend (1993) introduced a dynamic and stochastic model of decision-making under uncertainty, called *decision field theory (DFT)*.[2] About the same time, Usher and Zakay (1993) developed a dynamic connectionist type of model for multiattribute decision-making. Building on this previous work, Roe, Busemeyer, and Townsend (2001) extended DFT to choices among multiple alternatives and multiple attributes. Following soon after, numerous competing dynamic models of preferential choice have appeared (reviewed in the next chapter).

We begin by providing a comprehensive presentation of DFT. We do this for a couple of reasons. First, but not necessarily most important, it was developed by one of the authors of this book, and this is an opportunity to provide an overview of all the previous work. Second, and more important, DFT has been applied to a wider range of phenomena than most other dynamic preference models. Chapter 8 presents a review and summary of the other competing models, which we then compare with DFT.

7.1 Decision Field Theory for Binary Decisions under Uncertainty

We start with a simple binary decision under uncertainty type of problem, such as the one illustrated in table 7.1. Here, we use only three states of nature for simplicity, but the model applies to any arbitrary number of states of nature.

For example, if state of the world S_2 obtains, then choosing action A produces payoff r_{A2} and choosing action B produces payoff r_{B2}. The payoff, r_{ij}, is evaluated according to an affective system to produce some subjective reward or punishment value denoted as $u(r_{ij})$. This evaluation is analogous

Table 7.1
Payoff matrix for two courses of action and three states of nature.

	State S_1	State S_2	State S_3
Action A	r_{A1}	r_{A2}	r_{A3}
Action B	r_{B1}	r_{B2}	r_{B3}

to the utility function used in traditional decision theories as described in chapter 2. We assume that the decision-maker has learned something about the probabilities of the states of nature based on past experience, so the focus for this chapter is on choice rather than learning. Later in chapter 9, we discuss models that place a greater emphasis on learning.

According to DFT, the decision-maker's choice is based on the following sequential sampling process (see figure 7.1 for an illustration). In general, each action is associated with a preference state, which accumulates advantages or disadvantages for that action. The decision-maker begins with an initial preference state for each action—$P_A(0)$ for action A and $P_B(0)$

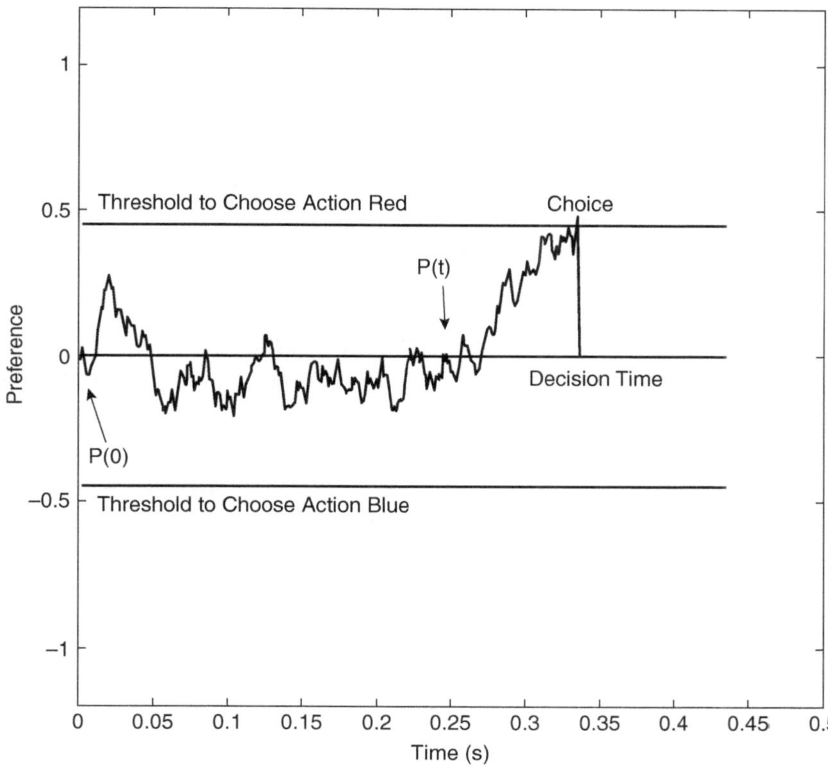

Figure 7.1
Sequential sampling model of binary choice. Time is on the horizontal axis, and preference is on the vertical axis. The upper and lower bounds represent thresholds. The trajectory represents the preference state evolving across time. The vertical line indicates where the process stops and makes a choice.

for action B. These initial preference states are based on past experience with each action or some initial status quo bias. For a binary choice, we are mainly interested in the difference between these two preference states, which is denoted as $P(0) = P_A(0) - P_B(0)$, which may be positive and favoring action A or negative and favoring action B. This initial state is then updated at each moment in time to produce a difference in preference states defined as $P(t) = P_A(t) - P_B(t)$ at time t. The process used to update $P(t)$ is described next.

At each step during the deliberation process, the decision-maker attends to one of the possible states of nature. If a state of nature is considered during some short period of time h, then the decision-maker evaluates each action according to the state of nature being considered during this time step. Attending to state S_i at time step t for a period of time h produces a valence difference equal to

$$v(t) = u(r_{\mathrm{Ai}}) - u(r_{\mathrm{Bi}}),$$

where $u(r_{\mathrm{Ai}})$, and $u(r_{\mathrm{Bi}})$ are the affective responses to the consideration of the payoffs for each action if state S_i obtains during the time period h. Then this valence is used to update the preference state as follows:

$$P(t+h) = (1 - \gamma \cdot h) \cdot P(t) + v(t), \tag{1}$$

where γ is a parameter in the model that controls primacy and recency effects in the accumulation process. If $0 < \gamma < 1$, then the accumulation process produces recency effects; if $\gamma < 0$, then the process produces primacy effects.

Attention switches from moment to moment to different states of nature to produce a sequence of valence increments, which are accumulated according to equation (1). The probability of attending to a state of nature within a period of time h during each time step is defined as $w(S_i) \cdot h$. We assume that $w(S_i)$ sums to 1 across states, forming a probability distribution over states.[3] The attention parameter $w(S_i)$ is analogous to a decision weight in the traditional utility theories discussed in chapter 2.

The accumulation process continues until the difference in preference state becomes sufficiently strong to exceed a threshold, P^*, in magnitude:

If $P(t) > +P^*$, then stop and choose A;

if $P(t) < -P^*$, then stop and choose B;

otherwise, continue sampling valences.

The basic idea is that the decision-maker needs to sequentially estimate the expected utility difference between actions using this sampling process, and the estimation process stops when the difference exceeds threshold.

7.2 Theoretical Parameters

Now we derive a number of theoretical parameters from DFT, which are used to predict choice probability and the distribution of response time from the DFT model. Defining $w(S_i) \cdot h$ as the probability of attending to a state at each time step, we can derive the mean valence at each time step as follows:

$$E[v(t)] = h \cdot \delta = h \cdot \sum_i w(S_i) \cdot [u(r_{iA}) - u(r_{iB})], \tag{2}$$

and the variance for each step as follows:

$$Var[v(t)] = \sigma^2 \cdot h = \sum_i h \cdot w(S_i) \cdot [u(r_{iA}) - u(r_{iB})]^2 - h^2 \cdot \delta^2. \tag{3}$$

For small h, we can ignore the second-order term $h^2\delta^2$. It is useful to define the change in preference as $dP(t+h) = P(t+h) - P(t)$, and rewrite equation (1) as a stochastic difference equation as follows:

$$dP(t+h) = -\gamma \cdot h \cdot P(t) + \delta \cdot h + d\varepsilon(t), \tag{4}$$

where $d\varepsilon(t) = v(t) - \delta \cdot h$, with mean equal to zero, $E[\varepsilon(t)] = 0$, and variance equal to $V[d\varepsilon(t)] = \sigma^2 \cdot h$. The mean increment in preference state, conditioned on the previous state, equals

$$E[dP(t+h) \mid P(t) = x] = (\delta - \gamma \cdot x) \cdot h, \tag{5}$$

and the variance of the increment equals $V[dP(t+h) \mid P(t) = x] = \sigma^2 \cdot h$.

Using these derivations, we define the standardized parameters as follows:

$d = (\delta / \sigma)$, standardized mean drift rate,

$z = (P(0) / \sigma)$, standardized starting value,

$\theta = (P^* / \sigma)$, standardized threshold.

As time step h approaches zero, this process can be approximated by a continuous time diffusion process called the Ornstein–Uhlenbeck (OU) process (Busemeyer & Townsend, 1992).[4] If we set the primacy–recency parameter to zero ($\gamma = 0$), then this process reduces to the Weiner process described in chapter 6.

Analogous to a signal detection type of model, the standardized mean drift rate, d, is determined by the ratio of the mean valence to the standard deviation of the valence. The standardized starting value, z, is assumed to be influenced by past experience and status quo effects. The threshold, θ, is assumed to be influenced by time pressure and the cost of time for sampling information. One very important feature of DFT is that the variance is *not* constant, but rather changes with the uncertain options according to equation (3).

DFT shares the same parameters as the weighted utility model described in chapter 2 for determining the mean valence δ. The variance σ^2 is computed using the same weights and values as the mean valence, and so this does not introduce any new parameters to be estimated from data. The growth-decay parameter γ is an additional parameter that is needed when there are recency or primacy effects. The threshold, θ, is an additional parameter that is needed to account for speed–accuracy trade-off effects.

Model predictions The mathematical formulas used to predict choice probability and the distribution of choice time for DFT (or for OU processes in general), are more complex than those for the simpler Weiner process. Therefore, we present these formulas in the first appendix of this chapter.[5] However, if we assume that $\gamma=0$ (i.e., we assume no primacy or recency effects), then we can use the simpler formulas for the choice probability and distribution of choice times obtained from the Weiner process as described in chapter 6. In the latter case, we simply insert the DFT parameters d, z, and θ into equations 1 and 2 of chapter 6. However, the equations given in the first appendix work equally well whether or not $\gamma=0$. They simply give the same answer as the Weiner process when $\gamma=0$.

Figure 7.2 shows an example of the predicted choice and response time distribution using the formulas given in the appendix. As can be seen in this graph, the distributions are very similar in shape to that produced by the diffusion (Wiener) model described in chapter 6. There is one important difference: DFT (and the OU process as well) do not necessarily predict equal mean response time conditioned on each alternative when the bias is zero. As can be seen in the insert in figure 7.2, the mean response time conditioned on choosing red (the more frequently chosen option) is faster than blue. Therefore, DFT does not have to assume variability in parameters to produce differences in mean response time. This does not mean that

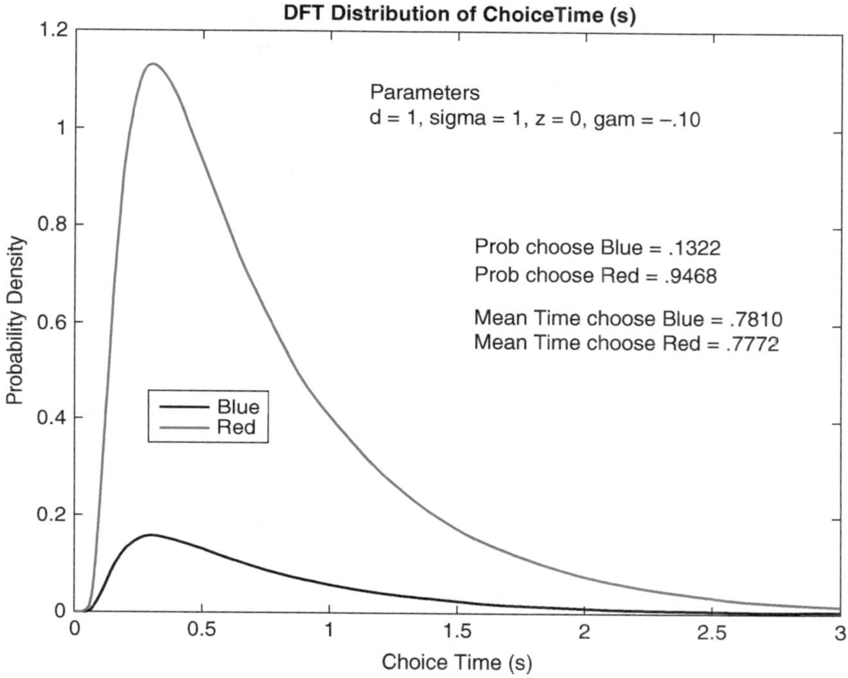

Figure 7.2
Choice and response time distribution predicted by DFT. The high curve shows the probability of stopping at time *t* and choosing red, and the lower curve shows the same for choosing blue. The high curve integrates to equal the choice probability given in the insert, which also shows the mean time to choose each action.

variability of parameters is not a good idea, but it does mean that there is more than one reason for differences in mean response time.

7.3 Applications to Binary Choice Findings

DFT has been shown to quantitatively fit choices between gambles better than popular choice models such as a probabilistic version of prospect theory (Rieskamp, 2008), and it also fits choices between consumer products better than another recently introduced cognitive model called the *proportional difference choice model* (Scheibehenne, Rieskamp, & Gonzalez-Vallejo, 2009). Next, we present some important qualitative predictions used to test the theory.

Variance effect An important principle of DFT is that the variance of the valence changes across choice pairs depending on the payoffs assigned to each action. Recall that the standardized mean drift rate parameter, $d=(\delta/\sigma)$, is a ratio of the mean valence divided by the standard deviation, σ, of the valence, and the latter changes depending on the payoffs of the actions. Considering table 7.1, we can manipulate the variance of action A by changing the range of payoffs for option A (i.e., changing the range across $\{r_{A1}, r_{A2}, r_{A3}\}$).

If we ignored the standard deviation (or assumed that it was constant across all pairs of actions), then this model would satisfy both strong stochastic transitivity and choice independence. Many authors have in fact assumed such a restrictive choice model (e.g., Stott, 2006). However, this is problematic because, as reviewed in chapter 3, there is extensive empirical evidence demonstrating the effect of heterogeneous variance across choice pairs (Blavatskyy, 2007; Hey & Orme, 1994; Erev & Barron, 2005; Busemeyer, 1985; Wilcox, 2011). As reviewed in chapter 3, this heterogeneity in variance produces systematic violations of both strong stochastic transitivity and choice independence. Ignoring heterogeneity in variance across conditions can distort other parameter estimates in a model (Wilcox, 2011). Therefore, it is important to include the change in variance across pairs of actions to account for these robust findings in the choice literature.

Consider, for example, the violation of binary choice independence reported by Busemeyer (1985) and discussed in chapter 3. Option A was a gamble producing normally distributed payoffs with a mean equal to zero and a standard deviation equal to 5 cents. Option B was a gamble producing normally distributed payoffs with a mean equal to zero and a standard deviation equal to 50 cents. Option C delivered a loss of 3 cents for sure, and option D delivered a gain of 3 cents for sure. The results found by Busemeyer (1985) were as follows:

$$p(A\,|\,\{A, C\}) = .88 > p(B\,|\,\{B, C\}) = .62,$$

but

$$p(A\,|\,\{A, D\}) = .11 < p(B\,|\,\{B, D\}) = .27.$$

The reversal of the two inequalities shown here indicates a violation of binary choice independence. DFT accounts for this finding using the following simple explanation: For the first inequality, the mean advantage is

$\delta=3$ for choosing *A* over *C*, and this is also the case for *B* over *C*. However, this mean advantage is easier to discriminate with the *A*, *C* choice pair than the *B*, *C* choice pair because the variance is much smaller for the *A*, *C* choice ($\sigma=5$) than for the *B*, *C* choice ($\sigma=50$) producing a larger positive ratio $d=+3/5$ for the *A*, *C* choice than the much smaller positive ratio $d=3/50$ for the *B*, *C* choice. Therefore, *A* is chosen over *C* more frequently than *B* over *C*. For the second inequality, the pattern reverses: Now there is a disadvantage equal to $\delta=-3$ for choosing *A* over *D* and the same disadvantage for choosing *B* over *D*; but now the ratio for *A* over *D*, $d=-3/5$, is much more negative than the ratio $d=-3/50$, for choosing *B* over *D*. Therefore, *A* is chosen over *D* much less often than *B* over *D*.

If the effect of variance is not taken into consideration in this example, then the results would have to be interpreted as a change in risk attitude depending on the value of the sure option. We would need to assume that the participants were risk seeking when the sure thing was positive, but they switched to risk aversion when the sure thing was negative. By allowing for heterogeneous variance, DFT provides an account that does not require changing risk attitude with the sign of the sure option.

Correlation effect There are two ways to manipulate the variance of the valence. One way, described previously, is to manipulate the variance of the payoffs produced by each course of action. Considering table 7.1, we can manipulate the range of payoffs for option *A* (i.e., the range across $\{r_{A1}, r_{A2}, r_{A3}\}$). Another way is to manipulate the correlation between the payoffs (the correlation between the pair of rows of table 7.1). The variance of the valence will be large if the payoffs produced by the pair of actions are negatively correlated rather than positively correlated.

For example, the payoffs shown in table 7.2 are positively correlated, whereas the payoffs shown in table 7.3 are negatively correlated. For simplicity, suppose that the three states of nature are equally likely. In the positively correlated case, a state that produces a win for action *A* also produces a win for action *B*, and another state that produces a loss for action *A* also produces a loss for action *B*. In the negatively correlated case, a state that produces a win for one action produce the opposite, a loss, for the other action.

Note that action *A* stochastically dominates action *B* in table 7.2, which is easily detected because the payoffs are positively correlated. Table 7.3 was constructed from table 7.2 simply by exchanging the columns of the payoffs

Table 7.2

Payoff matrix for positively correlated payoffs.

	State S_1	State S_2	State S_3
Action *A*	1.00$	0	–0.99$
Action *B*	0.99$	0	–1.00$

Table 7.3

Payoff matrix for negatively correlated payoffs.

	State S_1	State S_2	State S_3
Action *A**	1.00$	0	–0.99$
Action *B**	–1.00$	0	0.99$

for action *B* such that now the payoffs are negatively correlated. Assuming that the events are equally likely, then *A** also stochastically dominates *B**, but now it is difficult to detect.

According to the discrete time version of the DFT, when faced with the choice between *A* and *B* in table 7.2, each step in the sequential sampling process produces an increment in favor of action *A* (or no change at all), so action *A* is certain to be chosen over *B*. However, when faced with the choice between action *A** versus *B** in table 7.3, one-third of the time the increment strongly favors *A**, and one-third of the time the increment strongly favors *B**, and the remainder produced no difference. So, when faced with the *A** versus *B** choice pair, the valence swings from very positive to very negative, producing a large variance in the valence, and the choice probability is reduced to near chance. This correlation strongly influences the variance of the valence. For the positively correlated case, the variance of the valence is very small, and so the ratio $d=(\delta/\sigma)$ is large, producing a very high probability of choosing the dominant action A. For the negatively correlated case, the variance of the valence is very large, and so the ratio $d=(\delta/\sigma)$ is close to zero, producing a choice probability of close to .50. These predictions from DFT were confirmed in an experiment by Diederich and Busemeyer (1999), and more recently, more general effects of correlation between payoffs on choice were reported by Andraszewicz, Rieskamp, and Scheibehenne (2015).

Speed–accuracy trade-off The amount of time that a decision-maker allocates to making a choice affects the probability that an option will be chosen. Important decisions require more time, increasing the likelihood of choosing the best option, whereas unimportant decisions are made more quickly but often result in a bad choice. Some decision-makers tend to be more deliberative and take more time to decide, trying to make the right choice, whereas other decision-makers tend to be more impulsive and make quick decisions, caring less about mistakes. One of the unique features of DFT is its ability to account for the relation between speed and accuracy in preferential choice and risky decision-making.

The key idea is that the threshold bound, θ, controls the stopping time. When an important decision is to be made, then this threshold is set to a large magnitude, thus requiring a very high level of preference strength before making a decision. When an unimportant decision is to be made, or when the decision-maker is under time pressure to act quickly, then the threshold is set to a lower magnitude, allowing a weak preference toward making the decision. Thus, the threshold parameter provides a simple way to model the empirically observed relation between speed and accuracy of choice. More important, preference reversals can occur when the choice probability changes from above to below .50 (or vice versa) depending on time pressure. Figure 7.3 illustrates an example of the predictions for the speed–accuracy trade-off produced by increasing the threshold. In this example, the process starts out biased in favor of blue (e.g., blue is an initially preferred safe or status quo option). However, the mean drift rate favors red (e.g., red has a greater expected value than blue). Under short deadlines, the initial bias has a major influence, but as the threshold increases and the sampling process accumulates preference favoring the higher expected value option, the probability increases for the red option.

Busemeyer (1985) provided evidence for increases in choice probability for the higher expected value alternative with decreasing time pressure and longer deliberation times. Diederich (2003) provides even stronger evidence for preference reversals between risky and safe actions under time pressure.

Relation between choice and decision time One of the advantages of a dynamic preference model is the ability to model both choice and response times simultaneously using the same model and parameters. Accounting for both dependent variables with the same model provides an opportunity to conduct generalization tests in which model parameters are estimated from choice probability data, and then the same parameters are used to predict mean decision time. Petrusic & Jamieson (1978) originally found an

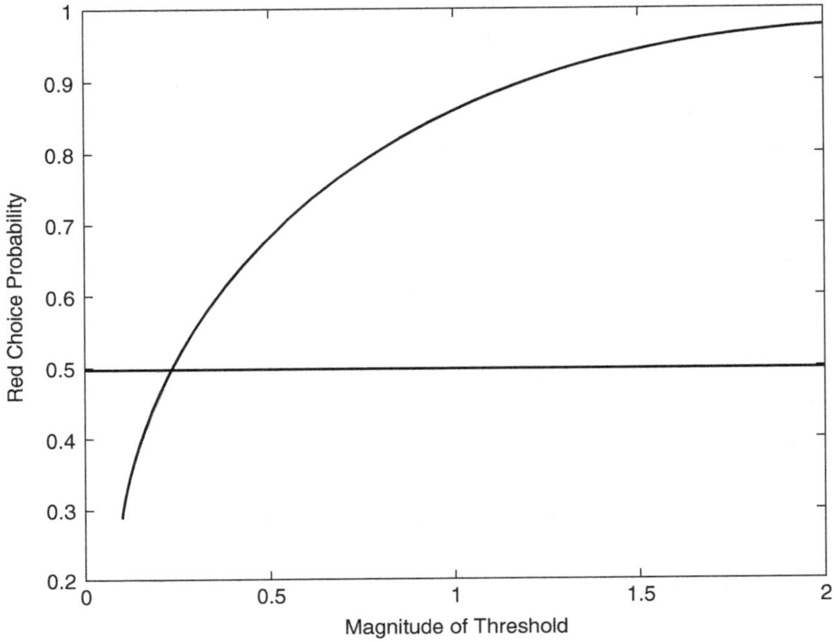

Figure 7.3
Speed–accuracy trade-off predicted by DFT. The horizontal axis represents the threshold magnitude, and the vertical axis represents the probability of choosing red. The process is initially biased toward blue, but as deliberation time lengthens, the probability for red increases.

inverse relation between the probability to choose an action and the mean time to choose that same action in preferential choice. Later, Busemeyer and Townsend (1993) conducted a large choice experiment to examine this relation. They fit the DFT model to their choice probabilities, and then used these same parameters to predict the inverse relation between the probability to choose an action and the mean time to choose that same action from theie experiment.

So far, we have focused on decisions under uncertainty, but DFT can also be applied to other types of decision tasks, such as the intertemporal choice task. In an intertemporal choice task, the decision-maker has a choice between a small payoff to be paid within a short time delay and a larger payoff to be paid after a longer time delay. Previous work on intertemporal choice never examined the time to make an intertemporal choice decision and its relationship with choice probability. Dai and Busemeyer (2014) examined this relation and found the same type of inverse relation between choice probability and decision time as with decisions made under uncertainty. DFT was successful in accurately predicting both the choice and decision times in this task. We also compared DFT to several competing intertemporal choice models, such as the hyperbolic discounting model, and showed that DFT provided more accurate predictions compared to the competing models.

Choice and certainty equivalents One of the most paradoxical findings in the decision-making literature is the systematic reversal of preference when it is measured by choice as opposed to price equivalence (Lichtenstein & Slovic, 1971; Grether & Plott, 1979; Tversky, Slovic, & Kahneman, 1990). Consider a choice between gamble *A* and gamble *B*. Gamble *A* delivers a relatively small payoff of $4, with a relatively high probability 35/36, or else −$1. Gamble *B* delivers a relatively large payoff of $16, but with a relatively low probability of 11/36, or else −$2. The well-known finding for this example is that gamble *A* is chosen more frequently than *B* in a binary choice, but the mean price equivalent given for gamble *B* is higher than that for *A*. Although preference reversals do not occur for all gambles, robust reversals do occur for gambles with features similar to these example gambles.

A popular account of this result, called the *contingent weighting model* (Tversky, Sattath, & Slovic, 1988), argues that when evaluating a gamble based on the trade-off between probability to win and amount to win, more weight is given to probability in the choice process, but more weight is given to price in the certainty equivalent price process. The problem with this is that it cannot account for the fact that preference reversals also occur between different pricing procedures (selling versus buying prices; Birnbaum & Zimmerman, 1998).

DFT uses the same model for choice as described here, and it provides a dynamic search process model to find a price equivalent to gambles (Johnson & Busemeyer, 2005). The basic idea is that when asked to find the price equivalent of a gamble, the decision-maker starts with an initial guess at the price equivalent (a candidate price) and then uses a preference evolution process to choose between the gamble versus the candidate price. If the preference process hits the bound to choose the gamble over the candidate price, then the price is too low and it is increased for the next guess; if the preference process hits the bound to choose the candidate price over the gamble, then the price is too high, and it is decreased. This search eventually stops and reports a final price equivalent by using the following stopping procedure: Each time a candidate prices is compared to the gamble, the preference state evolves across time back and forth between the candidate price and the gamble, and occasionally the preference state crosses zero, representing indifference. Each time the preference state crosses zero, there is a probability to stop the process to finally select and report the price.

This simple search mechanism is all that is needed to produce preference reversals without changing any weights for choice and price. Figure 7.4 illustrates the predictions of the DFT price equivalence model. The top panel shows the relative frequency distribution of prices predicted by the model for the P-bet, which is a distribution highly skewed to the left of $4 with a relatively small variance. The bottom panel shows the relative frequency distribution of prices predicted by the model for the D-bet, which is a distribution spread out around $4.50 with a relatively large variance.

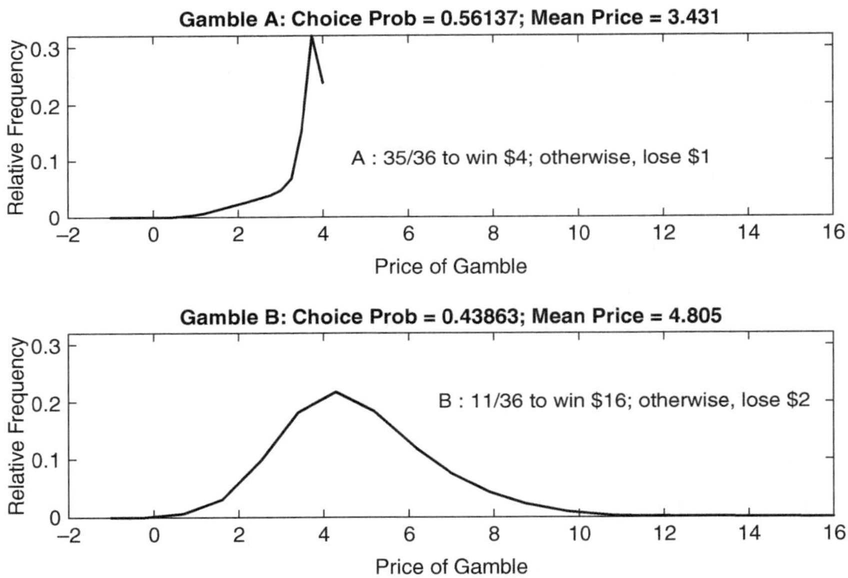

Figure 7.4
Relative frequency of certainty equivalents for gamble *A* (top) and gamble *B* (bottom), all produced by the DFT price equivalence model. The title above each panel describes the mean price and probability of choosing each gamble predicted by the model.

Also, this same model predicts preference reversals not only between choice and price, but also between buying and selling prices. Furthermore, a distinct advantage of this price search model is that it not only predicts choice probability and the mean price, but it also predicts the entire distribution of prices given to a gamble. In fact, the search model predicts that the variance of the price increases with the variance of the gamble, which is consistent with previous experiments (Bostic, Herrnstein, & Luce, 1990). More recently, Kvam and Busemeyer (2020) developed a more complete model that accounts not only for choices and prices, but also the time required to assign a price.

Risky decision paradoxes In chapter 2 on utility theory, we pointed out that various paradoxes were found, such as the Allais paradox and the Ellsberg paradox. These paradoxes pushed decision theorists to propose various types of nonlinear weighting functions to account for the results. Nonlinear decision weights are often used to account for these results. The probability of attending to a state corresponds to the decision weight in DFT. Initially, DFT did not specify any specific mechanism for the attention weights used in equations (2) and (3).

Johnson and Busemeyer (2016) addressed this problem by developing a mental simulation process that generates the decision weights for DFT. The basic idea is that a decision weight corresponds to the proportion of times, an outcome is predicted to occur when an action is mentally simulated. The mental simulation process was formalized as a Markov chain process, which transforms objective outcome probabilities into what correspond to decision weights. We won't go into the details here, but this mechanism (1) coherently explains many of the well-known paradoxes with a single set of parameters, including event splitting, which leads to violations of dominance; (2) generates a weighting function similar to Birnbaum's (2008b) tax model, which is one of the most successful utility models to date; and (3) provided better predictions compared to prospect theory in a generalization test that compared the model using extensive data sets on choice and pricing.

Another perhaps even simpler way for DFT to address these decision paradoxes was presented by Bhatia (2014). The basic idea is to allow two kinds of event sampling to occur during deliberation: During each time step, there is a probability π that the decision-maker attends to each state according to its described probability, and there is a probability $(1-\pi)$ that the decision-maker attends to each state with equal probability. If π is relatively high, then this mixed sampling procedure is able to reproduce the common-ratio, common-consequence, reflection, and event-splitting effects. Navarro-Martinez, Loomes, et al. (2018) later proposed a different kind of sequential sampling model to account for these paradoxes.

7.4 Multiple Alternative, Multiattribute Decision Field Theory

Roe, Busemeyer, and Townsend (2001) extended DFT beyond binary choice to a choice among multiple alternatives (i.e., more than two) described by multiple attributes (a consumer product with economic and quality attributes), which we called *multialternative decision field theory (MDFT)*. Figure 7.5 illustrates the basic ideas for MDFT.

Suppose that the decision-maker is presented with a set of m choice options (figure 7.5 uses 3 for an example). Each choice option can be evaluated on the basis of n attributes. For example, the decision-maker may evaluate three hotels on two attributes described by quality and economy, as illustrated in table 7.4. According to MDFT, an abstract attribute, such as economics, actually consists of a large number of specific features that need to be considered. Attention to an attribute at a moment in time is not just a simple read of a number from a table; instead, a number in a table serves as a cue for thinking about all the detailed implications (e.g., a hotel room might be rated with two stars on a website, but this can include many particular features, such as dirty sheets, poor service or noisy rooms).

The decision process starts with an $m \times 1$ state vector $P(0)$, where each coordinate corresponds to the preference for one of the options. In figure 7.5, there are three options, so $m=3$. As before, this initial state $P(0)$ reflects preferences for status quo options or biases from past experience. The initial state vector then evolves across time t by accumulating evaluations regarding the advantages and disadvantages of each option. After deliberating for time t, the cumulative evaluations evolve to a new $m \times 1$ state vector $P(t)$ according to the linear process

$$P(t+h) = (\mathrm{I} - \Gamma \cdot h) \cdot P(t) + V(t+h).$$

(In many previous applications of MDFT, it was assumed that $h=1$ for simplicity.) The $m \times m$ matrix Γ determines the nature of the feedback from the

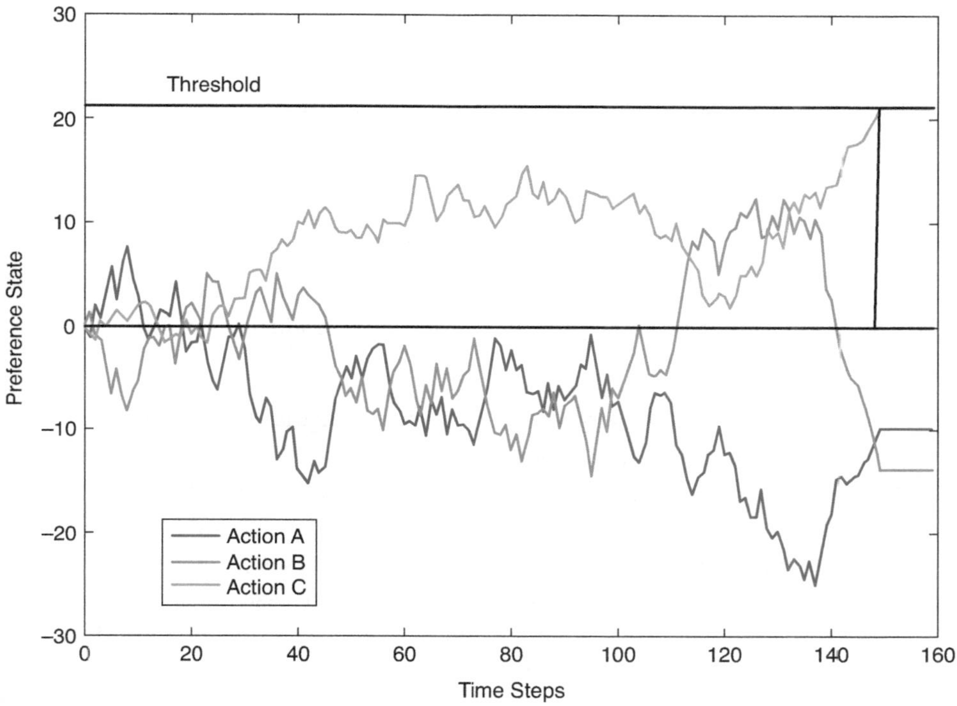

Figure 7.5

MDFT choice process for three alternatives. The horizontal axis represents time steps, the vertical axis represents preference for each action, and the flat horizontal line represents the decision threshold that has to be crossed by one of the preference states so that a choice occurs. In this case, option *C* first crosses the threshold at a time equal to 150 time steps and is chosen by the decision-maker. See color plate 6.

Table 7.4
Multialternative, multiattribute choice presentations.

	Quality Attribute	Economic Attribute
Option A	v_{AQ}	v_{AE}
Option B	v_{BQ}	v_{BE}
Option C	v_{CQ}	v_{CE}

previous state, which will be described in more detail later in this chapter. The valence input $V(t)$ is defined by the following product:

$$V(t) = C \cdot M \cdot W(t).$$

First, consider the $n \times 1$ vector $W(t)$, which represents the attention given to each attribute at time t. Table 7.4 has two attributes, and so we can assume temporarily that $n = 2$. This attention weight vector is assumed to change stochastically across time as the decision-maker's attention switches from one attribute to another. Next, consider the $m \times n$ matrix, M, which contains the subjective evaluation of each alternative (rows) for each attribute (columns). The product $M \cdot W(t)$ is analogous to a vector of weighted utilities, one for each row, except they change stochastically across time. Finally, the $m \times n$ matrix C compares each option against the others by subtracting the average of all the weighted utilities from the weighted utility of each option. This last step forces the valence vector $V(t)$ to sum to zero, so that an advantage for one option that is above average must result in a disadvantage for some other option that is below average.

The preference state continues to evolve across time until the preference states for one of the options gets sufficiently strong to cross a threshold. This stops the accumulation process, and the first option to cross the threshold is chosen. The time to make the decision is the time required for an option to hit the threshold. Note that this stopping rule is a satisficing rule because two options may be very close and hard to distinguish; nevertheless, the one that happens to reach a satisfactory level is chosen. There is no correct answer, and options only have to be satisfactory compared to the threshold to be chosen. As with the binary choice model, the threshold controls speed–accuracy trade-offs—increasing the threshold increases decision time but also increases the amount of deliberation and evaluation of options. Low thresholds produce fast decisions, but perhaps without sufficient evaluation of all the features related to an attribute. High thresholds provide more time to consider more features or attributes, but the decision-maker may have some external time or other pressures (e.g., a parent making a purchase at a store with tired and cranky children in tow), so decision time is costly.

At first, it seems that MDFT is inconsistent with DFT when $m=2$ because it seems that MDFT uses a two-dimensional process for binary choice, but the earlier description of DFT used a one-dimensional process. However, this apparent discrepancy between MDFT and DFT does not really occur, because the valence sums to unity, and so MDFT with *two* alternatives is actually a one-dimensional process, and MDFT reduces to be exactly the same as DFT for binary choice.

Stopping rules The stopping rule illustrated in figure 7.5 is called the *self-terminating* or *subject-controlled* stopping rule. The time to make a decision with this rule is a random variable that depends on the time to reach the threshold. Another kind of stopping rule is the fixed or externally controlled stopping rule. In this case, a fixed time period is given to make the decision, and it is assumed that the preference process continues without any threshold until the fixed time. At the fixed time, the process stops and the option with the maximum preference state is chosen. Although the self-terminating stopping rule is more common and natural, the mathematical prediction equations are simpler for the fixed stopping rule (see appendix in Roe et al., 2001, for the fixed stopping time mathematical equations).

7.5 Theoretical Parameters

Based on the process described thus far, we can theoretically derive the key parameters to make choice probability and decision time predictions. We assume that $W(t)$ is an independent stochastic process, with a mean equal to $\mathrm{E}[W(t)]=w\cdot h$, which represents a $n\times 1$ vector of probabilities to attend to an attribute during a step, and the variance-covariance matrix for $W(t)$ is defined as $\mathrm{Cov}(W(t))=\Psi\cdot h$. An assumption that we have used in the past (Roe et al., 2001) is that $\Psi=(\mathrm{diag}(w)-w\cdot w')$, like a multinomial model. Then the mean valence equals

$$\mathrm{E}[V(t)]=\mathrm{E}[C\cdot M\cdot W(t)]=C\cdot M\cdot \mathrm{E}[W(t)]=(C\cdot M\cdot w)\cdot h=\delta\cdot h,$$

where $\delta=C\cdot M\cdot w$. The variance–covariance for the valence equals

$$V[V(t)]=C\cdot M\cdot(\Psi\cdot h)\cdot M'\cdot C'=\Phi\cdot h.$$

Inserting these results, we obtain the following for the preference state at time $t+h$, given the preference state at time t:

$$E[P(t+h) \mid P(t)=X] = (I - \Gamma \cdot h) \cdot X + \delta \cdot h,$$

$$V(P(t+h) \mid P(t)=X) = \Phi \cdot h.$$

If we define the change in preference as $dP(t+h) = P(t+h) - P(t)$, then we can rewrite the basic accumulation process as the stochastic difference equation

$$dP(t+h) = -\Gamma \cdot h \cdot P(t) + \delta \cdot h + d\varepsilon(t+h),$$

where

$$d\varepsilon(t) = V(t) - \delta \cdot h = C \cdot M \cdot W(t) - C \cdot M \cdot w \cdot h = C \cdot M \cdot (W(t) - w \cdot h)$$

is a Brownian motion increment with a mean of zero and the variance-covariance matrix $V(d\varepsilon(t+h)) = \Phi \cdot h$.

So far, we have assumed that the decision-maker attends only to attributes that the researcher has manipulated or measured. However, the n attributes to which a decision-maker attends may include attributes that are not explicitly manipulated or measured by the researcher (e.g., the researcher measures economic and quality attributes, but the decision-maker also attends to other incidental factors, such as brand name and manufacturing country). These uncontrolled attributes contribute noise to the MDFT model. We assume that this noise has a mean of zero, but including this noise adds variance, so $V[d\varepsilon(t+h)] = \Phi \cdot h + C \cdot (\sigma_n^2 \cdot I \cdot h) \cdot C'$.

As the time step h approaches zero, this process can be approximated by a multivariate OU process (Busemeyer & Diederich, 2002). If we set the feedback parameter to zero ($\Gamma = 0$), then this process reduces to a multivariate Weiner process.[6]

Finally, we need to define the coefficients in the feedback matrix Γ. The positive feedback, diagonal values of Γ, represent self-feedback, and they serve as the memory for the preference state for each option. The negative feedback, off-diagonal values of Γ, represent lateral inhibition, and these are assumed to produce competition between options. The off-diagonal coefficient connecting one option to another is assumed to be negative, with a magnitude that decreases with the psychological distance between the two options. Hotaling, Busemeyer, and Li (2010) provide an explicit formula relating the lateral inhibition to the distance between options that uses only three parameters.

MDFT shares exactly the same parameters as a weighted utility model used in a traditional Thurstone model described in chapter 3 for determining the mean vector μ. The covariance matrix Φ is derived from the same parameters as the mean vector, except for one additional noise parameter. Three additional parameters are required to determine the feedback matrix Γ on the basis of the distance between the m options (using the method given in Hotaling et al., 2010). The Wiener process is obtained as a special case by setting $\Gamma = \mathbf{0}$, but this reduces the capability of accounting for context effects that we discuss next. The last parameter is the threshold bound when the self-terminating stopping rule is used.

Mathematical predictions The MDFT predictions for multiple choice options (i.e., more than two) are straightforward to derive for the fixed stopping rule (see appendix of Roe et al., 2001). For this reason, previous applications that carried out quantitative tests of MDFT used the fixed stopping time model. Essentially, the prediction equations are the same as used for the Thurstone model (described in chapter 2). For example, if there are three options, then we use a bivariate normal distribution to determine

$$p(A \mid \{A, B, C\}) = \Pr[P_A(T) - P_B(T) > 0 \cap P_A(T) - P_C(T) > 0],$$

$$p(B \mid \{A, B, C\}) = \Pr[P_B(T) - P_A(T) > 0 \cap P_B(T) - P_C(T) > 0], \text{ and}$$

$$p(C \mid \{A, B, C\}) = \Pr[P_C(T) - P_A(T) > 0 \cap P_C(T) - P_B(T) > 0].$$

However, we need to specify the mean preference vector $P(T)$ and variance-covariance matrix of the preference $V[P(T)]$ at the fixed stopping time T (T = number of time steps). The simple solution for the mean preference vector at time T equals

$$\mathrm{E}[P(T)] = \Gamma^{-1} \cdot [I - (I - h\Gamma)^T] \cdot \delta + (I - h\Gamma)^T \cdot P(0).$$

The variance-covariance matrix at time T equals

$$V[P(T)] = \Sigma_{j=0,T} (I - h\Gamma)^j \cdot \Phi \cdot [(I - h\Gamma)^j]^T.$$

The summation over time can be costly, with large time steps, in terms of computation time. Assuming that the eigenvalues of $(I - h\Gamma)$ are less than 1 in magnitude, then it is possible to work out a simpler, asymptotic solution (Berkowitsch, Scheibehenne, & Rieskamp, 2014). Recently, a tractable formula for the covariance matrix for arbitrary time T was derived by Hancock, Hess, and Choudhury (2018).

The mathematical formulas for predicting trinary choice and using the self-terminating stopping rule were developed by Busemeyer and Diederich

(2002). However, they are computationally intense to implement. It is simpler to use computer simulation in this case. The second appendix discusses the MATLAB program used to compute the predictions that are described next. Figure 7.5 was generated using this program.

7.6 Theoretical Predictions for Multialternative Decision Field Theory

In chapter 3, we reviewed several types of context effects on choice, which occur when the choice set affects the relations among choice probabilities of the options. There are many kinds of context effects, but we will focus on the similarity, attraction, and compromise effects, illustrated in figure 7.6, which are arguably the most important.

MDFT was designed to account for context effects, and it has been shown to quantitatively fit choices between consumer products better than popular probit and logit models when context effects are present (Berkowitsch et al., 2014; Hancock et al., 2018). However, in these previous applications,

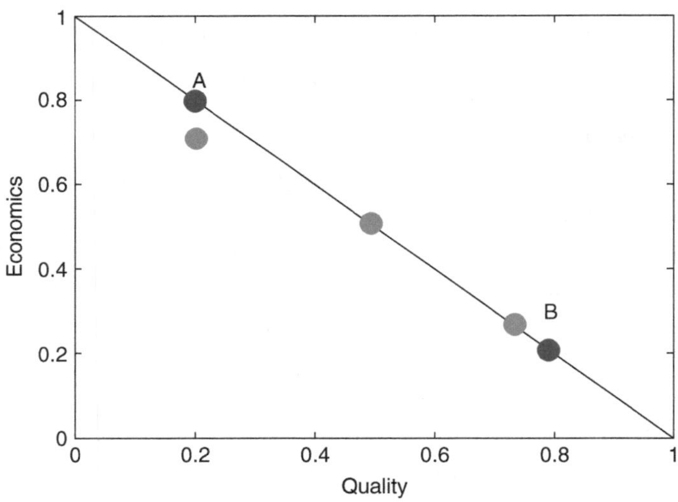

Figure 7.6
Choice options used to produce choice context effects. Option *A* is low on quality and high on economics, while option *B* is high on quality and low on economics. The option lying a bit above and to the left of *B* produces a similarity effect, the option a bit directly below *A* produces an attraction effect, and the option between *A* and *B* on the diagonal produces a compromise effect.

the fixed stopping time model was employed, because analytic solutions were available.

Below we describe predictions that MDFT makes for different types of context effects on choice using the self-terminating stopping rule. Using the self-teriminating stopping rule allows us to make predictions for both choice probabilities and mean time to make a decision. The simulation program used in the second appendix was used to compute these predictions.

Table 7.5 and figure 7.6 will be used to discuss these context effects and the model predictions. The data were generated using options described by two dimensions (e.g., quality and economics). Each row represents a choice set of three options. The columns labeled "Option A," "Option B," and "Option C" describe the values of the three options on the two attributes. The columns labeled "A," "B," and "C" display the predicted probability, and the column labeled "Time" lists the predicted time to make a decision (i.e., the number of steps needed to hit the bound). All the choice probabilities in table 7.5 were generated from MDFT using a single setting of parameters.[7] The model also generates predictions for binary choices, but except for the last three dominated rows, they are all equal to .50 and are not shown.

Similarity effect The pair of rows labeled "Sim" in the first column of table 7.5 illustrate the *similarity effect* because option C is similar to option A and different from option B. Adding C to a choice set that previously contained only A and B reduces the choice probability for A below B, even though the predicted binary choice between A and B is .50 in this case. In short, C, which is similar to and competitive with A, steals probability from A and leaves B intact. This kind of similarity effect was initially found by Tversky (1972). According to MDFT, the attention-switching mechanism is crucial for producing the similarity effect. The basic idea is that if the decision-maker attends to the second attribute, then B gets a unique advantage; but if the decision-maker attends to the first attribute, then A and C must split the advantage. If we eliminate the attention-switching process—that is, eliminate the variance $\Phi \cdot h$ from $V[d\varepsilon(t+h)] = \Phi \cdot h + C \cdot (\sigma_n^2 \cdot I \cdot h) \cdot C'$—then MDFT no longer predicts any similarity effect.

Attraction effect The pair of rows labeled "Attract" in the first column of table 7.5 illustrate the *attraction effect.* Option C is once again similar to option A, but now it is deficient or defective with respect to A. Adding

Table 7.5
Choice options designed to produce similarity (first two sets), compromise (second two sets), attraction (third two sets), and dominance (last three sets).

	Option A		Option B		Option C					
	Qual	Econ	Qual	Econ	Qual	Econ	A	B	C	Time
Sim	1	3	3	1	.9	3.1	.303	.403	.294	197
Sim	1	3	3	1	1.1	2.9	.286	.402	.311	190
Attract	1	3	3	1	.5	2.5	.775	.225	0	60
Attract	1	3	3	1	1.1	2.5	.567	.430	.003	97
Comp	1	3	3	1	1.75	2.25	.271	.316	.413	159
Comp	1	3	3	1	2	2	.301	.288	.411	153
Comp	1	3	3	1	2.25	1.75	.318	.276	.406	156
Dom	.5	.5	.7	.7	2	2	0	0	1	26
Dom	.75	.75	1	1	1.25	1.25	0	.072	.928	63
Dom	.5	.5	1	1	2	2	0	.021	.979	33

option *C* to the choice set now increases the probability of choosing option *A* because *C* makes *A* look good. The result is that adding *C* to the choice set containing *A* and *B* increases rather than decreases the probability of choosing *A*. In fact, the probability of choosing *A* rises above .50, which is the predicted binary choice probability between *A* and *B*, and so this effect produces a violation of regularity. This kind of attraction effect was initially found by Huber, Payne, and Puto (1982). According to MDFT, the lateral inhibition mechanism is crucial for producing the attraction effect. The defective option, *C*, attains a negative state, which feeds back through a negative lateral inhibition connection to produce a positive contrast effect on option *A*. If we eliminate the lateral inhibition by setting Γ equal to a diagonal matrix, then MDFT no longer predicts any attraction effect. The assumption that lateral inhibition is distance dependent is important here, because little or no lateral inhibition feeds back from the defective *C* to the more distant option, *B*.

Compromise effect The three rows labeled "Comp" in the first column of table 7.5 illustrate the *compromise effect*. Option *A* has a more extreme favorable value on the second attribute (economics) but low on the first, option *B* has a more extreme favorable value on the first attribute (quality) but low on second, and option *C* is moderate on both attributes. So, *C* is located between A and B on an indifference line with increasing quality and decreasing economics, which is why it is called a compromise. Note that a heuristic, such as a lexicographic or "take the best" heuristic, would always choose one of the extremes. But empirically, it has been found that there is a tendency to prefer the compromise in this case (Simonson, 1989). Even though all three options are often chosen equally in a binary choice, the compromise emerges as the preferred option in a trinary choice. According to MDFT, the lateral inhibition again is critical for producing the compromise effect. The basic idea is that although all three options have equal mean value, on those occasions that C happens to grow stronger by chance, it will inhibit both A and B, which gives it the advantage. Once again, if we eliminate the lateral inhibition by setting Γ equal to a diagonal matrix, then MDFT no longer predicts any compromise effect, and the effect also depends on the lateral inhibition distance assumption.

Dominance effect The last three rows labeled "Dom" in table 7.5 show the predictions of MDFT, using the same parameters, for situations in which option C dominates both options A and B. The first of these three illustrates a case of a fairly large dominance and, as expected, the model predicts that the dominant option is always chosen. In the second case, dominance is still present, but the differences are much smaller. The noise from irrelevant dimensions allows a dominated option to be chosen. If the noise from irrelevant options is set to zero, then the dominant option is always chosen. In fact, most experiments report some small proportion of trials in which participants choose a dominated option, and this could be due to attention to irrelevant options.

Decision time predictions The last column of table 7.5 shows the mean number of samples that reached the threshold. Note that the option set used to produce the similarity effect is predicted to take the most time; the attraction and compromise sets take the second most; and the dominance sets produce the fastest choices. These decision time predictions are just beginning to be examined more carefully by researchers, and they are discussed in the next chapter when we compare models with respect to these predictions.

The dynamic feature of MDFT provides a strong a priori prediction that has been tested. Both the attraction and compromise effects require the lateral inhibition mechanism, and the effect of this mechanism takes time to

Table 7.6
Choice options designed to produce similarity, compromise, attraction, and dominance using a relatively low threshold.

	Option A		Option B		Option C		A	B	C	Time
Sim	1	3	3	1	.9	3.1	.240	.499	.261	37
Sim	1	3	3	1	1.1	2.9	.256	.491	.253	38
Attract	1	3	3	1	.5	2.5	.501	.498	.001	22
Attract	1	3	3	1	1.1	2.5	.431	.529	.040	31
Comp	1	3	3	1	1.75	2.25	.341	.414	.246	39
Comp	1	3	3	1	2	2	.382	.379	.239	40
Comp	1	3	3	1	2.25	1.75	.414	.339	.247	40
Dom	.5	.5	.7	.7	2	2	0	0	1	11
Dom	.75	.75	1	1	1.25	1.25	.002	.138	.859	30
Dom	.5	.5	1	1	2	2	0	.008	.992	12

build up. Thus, MDFT predicts that seemingly irrational behavior, such as violations of regularity, increase with longer deliberation times. This prediction is counter to many heuristic choice accounts of these context effects because, presumably, fast and frugal heuristic rules are used more often when there is time pressure. Table 7.6 shows the predictions from MDFT using the same parameters as table 7.5, except that the threshold is reduced by half: The time to make a decision is much shorter; the similarity effects remain intact; the compromise and attraction effects are eliminated; and there is an increase of violations of dominance for the case that has small differences. Pettibone (2012) empirically tested these predictions for compromise and attraction effects by manipulating the time pressure to make a decision, and the results revealed decreased effect sizes with increasing time pressure, which agrees with the predictions of MDFT. However, Trueblood, Brown, and Heathcote (2014) empirically found reduced similarity effects under time pressure, but those results were obtained using an inference (e.g., infer guilt) rather than a preferential (e.g., choose a car) choice task.

7.7 Multistage Extension of Decision Field Theory

Diederich (2003a,b; Diederich & Trueblood, 2018) introduced an important new feature to MDFT concerning the attention-switching mechanism. Previously, it was assumed that attention switches fairly rapidly between attributes. This type of switching process can be approximated by a multinomial process, which produced a single, stationary mean valence that was the result of averaging over rapid switches in attention during deliberation. Diederich (2003a,b) proposed an alternative idea that attention remains focused on one attribute for an extensive period of time, and then switches to another attribute for another relatively long period of time, and possibly switches back to an earlier attribute later for some period of time, and so on. The simplest case occurs when there is a single switch from the first attribute to the second. This process is illustrated in figure 7.7, which shows the preference state accumulating in one direction favoring the first attribute that is processed (e.g., economics) and it almost but not quite hits the bound; but then it begins accumulating in a different direction after the switch to the second attribute (e.g., quality). The decision process could stop to produce a choice on the first attribute with a lower bound or continue on and process the second attribute with a higher bound.

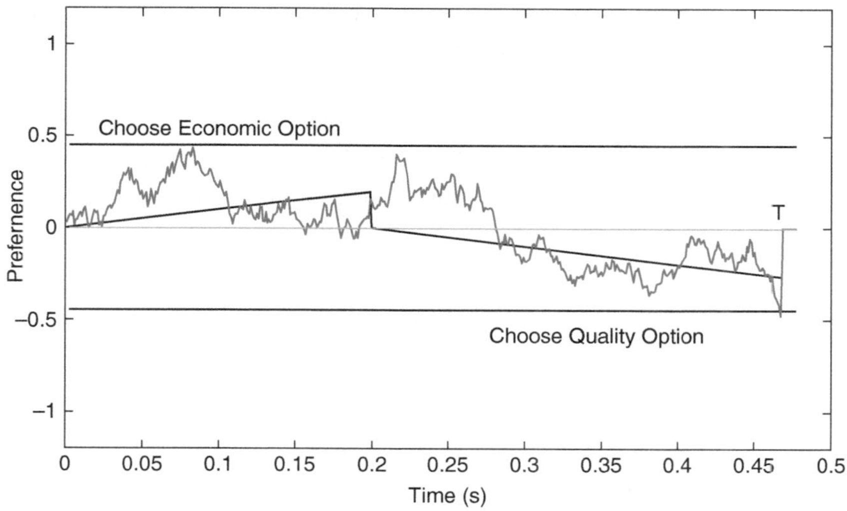

Figure 7.7
Illustration of the multistage model. The graph shows a binary choice between the high economic option and the high quality option. Initially, the process attends to the economic attribute and almost hits the threshold, but then a switch in attention occurs, and preference accumulates toward the choice of the high economic option. Eventually, the relative preference reaches the threshold for the high-quality option, which is chosen at time *T*.

The multistage model assumes that attention switches from one attribute to another according to a Markov process that uses a high probability of remaining within the attribute under consideration and a low probability to switch to a different attribute. Mathematical methods for computing the choice probability and the distribution of decision time were developed using the methods described in the first appendix (Busemeyer & Diederich, 2002).

There are several advantages to this multistage model. One is that it provides a better representation of preference reversals under time pressure. To achieve this with the earlier DFT model, it was necessary to assume that the process started with a bias toward one option, and then gradually accumulated evidence for the other option. The multistage model accounts for preference reversals with time pressure by assuming that under time pressure, the thresholds are decreased, which only allows time to process the first attribute; but without time pressure, the bounds can increase, allowing time to process the second attribute. Another important advantage of the multistage model concerns its predictions for mean decision time conditioned on the chosen option. If the attribute with the larger mean valence is processed first, then mean decision time to choose the more favorable option is shorter than the less favorable option; but if the attribute with the smaller mean valence is processed first, then the mean decision time to choose the more favorable option is longer than with the less favorable option. Using this model, unlike the drift diffusion model discussed in chapter 6, it is not necessary to employ parameter variability to account for differences in mean response time conditioned on a choice.

7.8 Decision Field Theory with Dynamically Changing Motivational Values

In all the previous applications discussed in this chapter, the main source of dynamics was generated from attention shifting, represented by the stochastic weight vector $W(t)$. In particular, these applications of DFT basically assumed that the affective values in the M matrix (e.g., as in table 7.4) were constant. Busemeyer, Townsend, and Stout (2002) proposed an additional dynamic process to describe how the affective values $M(t)$ can change across choice occasions driven by dynamically changing needs. For example, consider a student allocating time to work and leisure. The student might need to achieve a goal on a research project and work hard for many days. But

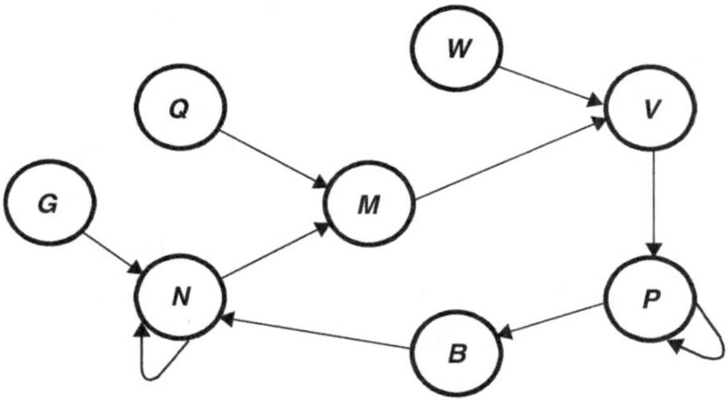

Figure 7.8
Diagram of directed influences for *goals* for each attribute (G), *need* for each attribute (N), *quantity* of each attribute produced by each choice (Q), *motivational* values for each choice on each attribute (M), *valence* for each action (V), *preference* for each option (P), and choice behavior (B).

after this hard work, the student may feel that she has made good progress on the project and needs some rest and recreation. Then, after some days of rest and recreation, the need to get back to work may build up again.

The formal mathematical description of the model is provided in Busemeyer, Townsend, and Stout (2002), and here we will only give a brief intuitive summary. Figure 7.8 illustrates the basic ideas behind the model for dynamically changing motivational values $M(t)$. The current needs (N) for each attribute combine with the quantities (Q) produced by each choice on each attribute to determine the motivational values (M) at each time point (the arrows from N and Q pointing into M indicate that M is a function of the two factors). The current motivational values (M) combine with the attention weights (W) to form the valence (V). The input valence (V) combines with the previous preference state (P) to form the new preference state across time. The preference state then determines the choice that is made, and this choice produces behavioral consequences (B) that feedback to change the achievement of needs (N). Finally, the behavioral consequences combine with previous needs and possible new outside goals (G) to form the new state of needs. Wang, Busemeyer, and Lang (2006) and Wang (2014) successfully applied this model to account for channel-changing behavior (i.e., the choices of each channel, time spent making a choice on a channel) when watching various types of news stations over time.

7.9 Applications of Decision Field Theory to the Sports, Consumer, Computer Science, and Engineering Fields

Although DFT was originally developed to account for experimental findings in decision-making, several extensions and applications of the basic ideas have appeared outside of experimental decision research.

One of the first applications was to decision-making in sports (Raab, 2002). These types of decisions are fundamentally dynamic. Sequential sampling models are motivated by their correspondence with the dynamic, variable processes that characterize decision-making in sports. Johnson (2006) provided a detailed description of how these process models can be applied to sports decisions, as well as incorporating other critical components, such as online option generation.

Gao and Lee (2006) and Gao, Lee, and Zhang (2006) extended DFT to describe reliance on automation in supervisory control situations. The model describes the iterated decision process and the evolution of preference for automatic and manual control. The extended DFT model captures several consistent empirical findings, such as the inertia of trust and the nonlinear characteristics of trust and reliance.

Lee, Son, and Jin (2008, 2010) developed an integrated Belief-Desire-Intention (BDI) modeling framework for human decision-making and planning for evacuation scenarios, whose submodules are based on the Bayesian belief network (BBN), DFT, and the probabilistic depth first search (PDFS) technique. A key novelty of this model was its ability to represent both the human decision-making and decision-planning functions in the unified framework. Simulations from the model were used to test the impact of several factors, such as demographics, number of police officers, and information sharing, on evacuation performance, including evacuation time and number of casualties. More recently, Lee and Son (2020) proposed combining the DeGroot (1974) consensus model with DFT to build a social-learning model for long-term decision-making processes in social networks.

Sun and Helie (2013) incorporated DFT into CLARION, a special module in a component of the non-action-centered subsystem (NACS) within the general architecture. The DFT network in the bottom level of the NACS enhances both the decision-making capabilities of CLARION and the range of the phenomena to which DFT can be applied.

Hancock and colleagues (2018) applied DFT to a large data set on transportation preferences and compared it to rival random utility models (e.g., multinomial logit models) commonly used in this field. They also extended the theory by including sociodemographic variables and ratings on multiattribute scales. DFT achieved significantly better fit in estimation and forecasting for data sets based on stated preferences for route choice.

Farahmand (2017b) used MDFT to model context effects obtained from the decisions of 25,646 domain name registrants. These findings have implications for Internet governance, as well as organizations that measure similarity and competition in domain name registration. In a more recent work, Farahmand (2017b) outlines how to apply DFT to privacy decision-making.

7.10 Concluding Comments

Decisions take time, and the time it takes to make a decision is an important factor for determining choice. The sequential sampling model in chapter 6 provided an extension of the static signal detection in chapter 5 that can account for both choice and decision time for evidence-based choice tasks. The sequential sampling model presented in this chapter provides an extension of the random utility models in chapter 3 which accounts for both choice and decision time for preference-based choice tasks.

In this chapter, we described DFT, one application of sequential sampling models to preferential choice. We focused on DFT here because it was one of the first to apply sequential sampling models to preferential or value-based choices, and it also has generated the largest number of applications to date. We showed how DFT can account for a wide range of findings, including results from decisions made under uncertainty as well as multiattribute, multialternative choice problems. However, a number of competing dynamic models of preference have appeared that also account for some of the same findings. The next chapter reviews the newer competing models and compares them with DFT.

7.A.1 Mathematical Formulas for Binary Choice Problem

This appendix derives formulas for computing the choice probability and distribution of stopping times for the binary choice version of DFT. It turns out that it is difficult to use the mathematical methods described in chapter 6 unless we assume that $\gamma = 0$, which reduces the OU model to a Weiner process. Therefore, we will use a different method to derive these probabilities in this chapter (see Diederich and Busemeyer, 2003). The method used in this chapter could have been used to derive the predictions used in chapter 6. If one uses the methods in this chapter and sets $\gamma=0$, then the same results are obtained as those produced by the equations used in chapter 6 (see the appendix of chapter 4 in Diederich and Busemeyer (2003) for the connection between the two methods when $\gamma=0$).

To derive the choice probabilities and distribution of choice times, we will employ a Markov chain approximation method (Bhattacharyia & Waymire, 1990, chapter 4). First, we divide the continuous preference interval $[-\theta, +\theta]$ into discrete levels separated by the width $\Delta=\sqrt{h}\cdot\sigma$. This produces a sequence of $2m+1$ discrete preference levels, where m equals the integer portion of θ/Δ:

$$(-\theta, -\theta+\Delta, -\theta+2\Delta, \ldots, -2\Delta, -\Delta, 0, +\Delta, +2\Delta, \ldots., \theta-2\Delta, \theta-\Delta, +\theta). \tag{5}$$

For the Markov chain approximation, we assume that during each time step h, there is a probability $p(x)$ to take a step up from preference level x to level $x+\Delta$, and a probability $1-p(x)$ to take a step down from preference level x to level $x-\Delta$. The probability to take a step up is defined as follows (the reason is described a bit later in this appendix):

$$p(x)=\frac{1}{2}\left(1+\frac{(\delta-\gamma\cdot x)\cdot\sqrt{h}}{\sigma}\right), \tag{6}$$

The probability to take a step down is defined as $1-p(x)$.

According to the Markov chain approximation, the preference level at time $t+h$ is defined by

$$X(t+h)=X(t)+\Delta(t),$$

where $\Delta(t)=+\Delta$ occurs with probability $p(X(t))$ and $\Delta(t)=-\Delta$ occurs with probability $1-p(X(t))$.

The mean change in preference level at a time step t equals (the far right-hand side uses the mean of the increment from equation 1 described earlier in this chapter)

$$E[\Delta(t)\,|\,X(t)=x]=p(x)\,\Delta-(1-p(x))\Delta=(2p(x)-1)\Delta=(\delta-\gamma\cdot x)\cdot h,$$

which is identical to the mean increment $E[dP(t+h)\,|\,P(t)=x]$ for the OU process . The variance of the change in preference level equals

$$V[\Delta(t)\,|\,X(t)=x]=p(x)\cdot(+\Delta)^2+(1-p(x))\cdot(-\Delta)^2-(\delta-\gamma\cdot x)^2\cdot h^2=\Delta^2=\sigma^2 h,$$

which is identical to the variance of the increment for the OU process (the extra terms with h^2 can be ignored for small values of h). For small time steps,

h, these definitions for transition probabilities (transiting up or down a level) produce a process that closely mimics the OU process. This is the reason for choosing these definitions of the transition probabilities in equation (6).

Using these definitions, we can define a $2m+1 \times 2m+1$ transition matrix T with rows and columns representing the sequence of preference levels (see expression 5 earlier), and with transition probability matrix elements, $T(y, x)$ to row y from column x, defined by $p(x)$ as follows:

$T(x+\Delta, x) = p(x)$ for $-\theta < x < \theta$, representing a transition from x to $x+\Delta$,

$T(x-\Delta, x) = 1-p(x)$ for $-\theta < x < \theta$, representing a transition from x to $x-\Delta$,

and with boundary conditions,

$T(-\theta, -\theta) = 1 = T(+\theta, +\theta)$ for the two absorbing states.

Define a $2m-1 \times 2m-1$ matrix Q as the transient state matrix containing preference levels $-\theta < x < \theta$, obtained by deleting the first and last columns and the first and last rows of T. Define R_A as a $1 \times 2m-1$ row vector containing the bottom row of T corresponding to the absorbing state θ, except that the first and last columns are deleted so that it contains only columns corresponding to $-\theta < x < \theta$. Define R_B as a $1 \times 2m-1$ row vector containing the top row of T corresponding to the absorbing state $-\theta$, except that the first and last columns are deleted so that it contains only columns corresponding to $-\theta < x < \theta$. Finally, define Z as a $2m-1 \times 1$ column vector containing the distribution over the starting positions for transient states $-\theta < x < \theta$. In particular, if we assume that the process starts at preference level z equal to the integer portion of z/Δ, where z is the standardized initial state of the OU process, then Z will have a 1 located in the row corresponding to z and zeros everywhere else. But this is not necessary; instead, we can assume an approximate normal distribution that is centered around z.

Using these definitions, we can derive the following predictions from the Markov chain approximation to the OU process. The probability to choose each action equals

$p(A) = R_A \cdot (I-Q)^{-1} \cdot Z$, $p(B) = R_B \cdot (I-Q)^{-1} \cdot Z$.

The mean time to make each choice equals

$E[T \mid A] = h \cdot R_A \cdot (I-Q)^{-2} \cdot Z/p(A)$, $E[T \mid B] = h \cdot R_B \cdot (I-Q)^{-2} \cdot Z/p(B)$.

The probability to make a choice at time $t = (n+1) \cdot h$ equals

$p(A \text{ at time } t) = R_A \cdot Q^n \cdot Z$, $\Pr[B \text{ at time } t] = R_B \cdot Q^n \cdot Z$.

The matix Q can be diagonalized as $Q = V \cdot \Lambda \cdot V^{-1}$, where V is the matrix of eigenvectors, and Λ is the matrix of eigenvalues. Then we can efficiently compute

$$Q^n = V \cdot \Lambda^n \cdot V.$$

These computations are simple to perform using mathematical programming languages such as MATLAB, R, Mathematica, and Python. MATLAB programs to compute these predictions are given at https://github.com/jbusemeyChoiceModeling.git.

7.A.2 Multialternative Choice with Self-Terminating Stopping Rule

The key program function in the set of functions for generating and fitting MDFT using the self-terminating stopping rule is given here. There are several other programs that are needed and the website https://github.com/jbusemeyChoiceModeling.git contains all of the program functions used to fit the model to data.

We set the time per step $h = 1$ simplicity. The program computes $N = 10{,}000$ simulations. (Note that these values can be adjusted.) The remaining lines should not need to be changed.

Inside the loop, we first generate a random weight vector w that contains a 1 in a single row and zeros otherwise. This represents picking an attribute. This used to update the preference state. The loop continues until the threshold is exceeded. The probability of picking each option is computed at the end. This simulation code can be accelerated by transforming it into machine code (which is also accessible on the website):

```
function [P3, T]=simMDF(S3,C3,M3,w,theta1,sig2,N)
%function [P3 T]=simMDF(S3,C3,M3,W,theta1,sig2,N)

nd=size(M3,1); % no. of alt's
V3=C3*M3*w; % mean valence
P3=zeros(1,nd); % counter
T=0; % No. Steps to reach decision

for n=1:N
  B=0; t=0; P=zeros(nd,1);
  while (B < theta1)
```

```
    W=w(1) > rand; W=[W ; (1-W)];
    E3=C3*M3*(W-w)+sig2*C3*randn(nd,1);
    P=P+S3*P+V3+E3;
    [B,Ind]=max(P) ;
    t=t+1;
% [n t]
  end
  P3=P3+[Ind ==1 Ind==2 Ind==3];
  T=T+t;
end

P3=P3./N;
T=T./N;
```

8 New Dynamic Preference Models

This chapter reviews and compares several new stochastic (dynamic and probabilistic) models of preferential choice that have appeared over the past thirty years (for reviews, see Wedell, 2015, and Busemeyer et al., 2019). All these models share the idea that each option is associated with an accumulator, denoted as $P_i(t)$ for the preference state of option i accumulated up to time t. For the self-terminating stopping rule, each accumulator races across time until one crosses a threshold, denoted as P^*, which determines the choice and decision time. For the fixed stopping rule, the accumulators continue up to the deadline, at which time the maximum is chosen. In this discussion, we summarize and compare all the models.

In principle, these stochastic choice models can be applied to a wide variety of decisions, including (1) decisions under uncertainty, with each action described by a value that is obtained if a state of nature is realized; (2) multiattribute choice problems, with each option described by values on attributes; or (3) intertemporal decisions, with each option described by a value that occurs at some future time. Some of the models reviewed here have been applied to all of these types of decisions, but others have been applied to only one of these types (e.g., multiattribute choice).

To facilitate comparisons among models, we use the same symbols for variables and parameters that share a similar function. Also, to keep it simple, we set $h=1$ for the time period of each step for all the models. For all the models, we will use P_i to denote the preference state for alternative i; we denote v_{ij} as the objective value of alternative i on attribute j; $u(x)$ denotes an increasing utility function; and w_j denotes the probability of attending to attribute j (or the probability of attending to state of nature j).

8.1 Seven Competing Dynamic Preference Models

Multialternative decision field theory (MDFT) again As described in chapter 7, Busemeyer and Townsend (1993) initially developed DFT for decisions under uncertainty, and later they expanded it to multiattribute choices with multi-alternative decision field theory (MDFT) (Roe, Busemeyer, & Townsend, 2001). First, it is assumed that attention switches across time from one attribute to another. The stochastic weight, $W_j(t)$, represents the attention weight given to attribute j at time t. At each time step, each attribute has a fixed probability of being selected, $w_j = \mathrm{E}[W_j(t)]$.

Second, a random weighted utility is computed for each action at time t:

$$U_i(t) = \sum_j W_j(t) \cdot u_{ij},$$

where u_{ij} represents the subjective evaluation of the value *vij* of option i on attribute j. The next step is to compute the valence of each action by comparing it with other actions:

$$V_i(t) = U_i(t) - \frac{1}{n-1} \sum_k U_k(t)$$

Then the preference state is updated according to the following stochastic difference equation:

$$P_i(t+1) = (1-\gamma) \cdot P_i(t) - \sum_{k \neq i} \gamma_{ik} \cdot P_k(t) + V_i(t+1) + \varepsilon(t+1).$$

The last term, $\varepsilon(t+1)$, is a noise term produced by attention to irrelevant dimensions. The feedback parameter γ accumulates the previous evaluations, and the parameters γ_{ij} for $i \neq j$, are negative lateral inhibitory links whose strengths are a function of the distance between alternatives (see Hotaling et al., 2010, for the specific function).

For binary choice, the preference states sum to zero, and so the process is one dimensional. This allows the derivation of mathematical formulas for the binary choice and decision time distributions for self-terminating tasks (see chapter 7). For choices with multiple options (i.e., more than two), mathematical formulas are available for fixed time decisions, but computer simulation is needed for self-terminating decisions.

Leaky Competing Accumulaor (LCA) model As described in chapter 6, the Leaky Competing Accumulator (LCA) model was originally developed by Usher and McClelland (2001) for evidence accumulation tasks such as signal detection. Later, Usher and McClelland (2004) extended the model to preferential choice (see also Tsetsos, Usher, & Chater, 2010). This model is fairly similar to MDFT, but with some critical exceptions. First, unlike MDFT, preference states can never go negative (because they interpret the state as a neural firing rate), which introduces nonlinearity into the dynamics. Second, contrary to MDFT's assumption that lateral inhibition is distant dependent, the LCA model assumes that lateral inhibition is constant across links. Together, these factors prevent the model from accounting for attraction and compromise effects without adding a new mechanism. The new mechanism introduced is loss aversion, a concept originally proposed by Tversky and Simonson (1993), which is not assumed by MDFT.

Like MDFT, the LCA model assumes that attention switches from one attribute to another. At each time step, each attribute has some fixed probability of being selected. Again, define $W_j(t)$ as the stochastic weight at time t; and at each time step, each attribute has a fixed probability of being selected, $w_j = \mathrm{E}[W_j(t)]$.

If attention is focused on attribute j at time t, then the person evaluates a nonlinear transformation of the advantage and/or disadvantage of every pair of options at that moment. The evaluation for the pair (i,k) at time t is

$$V_{ik}(t) = \sum_j W_j(t) \cdot u(v_{ij} - v_{kj}),$$

where the utility function $u(x)$ incorporates loss aversion [e.g., $u(x) = x^\alpha$ for $x > 0$, $u(x) = -\lambda|x^\alpha|$ for $x < 0$, and $\lambda > 1$].

The second step is to compute the net input of an option:

$$I_i(t) = I_0 + \sum_{k \neq i} V_{ik}(t).$$

This drives the accumulation process:

$$P_i(t+1) = (1-\gamma) \cdot P_i(t) - \sum_{k \neq i} \beta \cdot P_k(t) + I_i(t+1) + \varepsilon(t+1),$$

where $\varepsilon(t+1)$ is a noise term and β is the constant lateral inhibition. However, unlike MDFT, the preference state $P_i(t+1)$ is never permitted to go below zero in the LCA model. If the update computed for $P_i(t+1)$ falls below zero, then $P_i(t+1)$ is truncated to zero. Also, unlike MDFT, the preference states do not sum to zero for two alternatives, so the dimension remains the same as the number of options. This makes the model more complex, and computer simulation is needed. Even binary choices must be simulated in a two-dimensional plane.

Decision by sampling (DBS) Stewart, Chater, and Brown (2006) proposed a process for determining subjective values called *decision by sampling (DBS)*. The basic idea is that the subjective value x of an attribute (e.g., money, probability, or delay) equals the relative frequency that the target value x is larger than other members in memory of a comparison sample. The sample in memory could come from the local context (values of the attribute within the presented choice set) or from long-term memory (values experienced in the real world). The idea of using rank to determine value goes back to Parducci (1965). It also implements efficient coding (Bhui & Gershman, 2018).

Stewart and Simpson (2008), and later Noguchi and Stewart (2018), extended DBS by including a sequential sampling process for making choices. A counter $P_i(t)$ is assigned to each option. In each step, an option i is selected, and a target value is selected for option i, which is compared to a comparison value. Favorable comparisons on step t for option i increase the counter $P_i(t)$; otherwise, the counter remains unchanged.

Another important feature of the DBS model concerns the stopping rule. Most of the models use a satisficing type of self-terminating rule: stop as soon as a preference state exceeds the threshold P^*. But the DBS model uses the next best stopping rule: stop as soon as the first-ranked (maximum) preference state exceeds the second-ranked preference state by a threshold P^*.

The probability of selecting a target value on option i for attribute j is an increasing function of the similarity of the value of option i on attribute j to the set of other values on attribute j that are under consideration at that time. The set of other values under consideration is determined by a probabilistic selection from the values on the attribute of the other options presented in the choice set or a long-term memory sample.

The comparison process is simple for some attributes, such as money. If the target is larger than the comparison by a particular threshold, counter $P_i(t)$ increases by 1; otherwise, it remains unchanged. It is more complex for other attributes, such as probability. The sign (i.e., the "goodness" or "badness" of the value associated with a target probability) needs to be taken into consideration. If the value associated with a target probability is "relatively good," then a positive difference in probability that exceeds the threshold increments the counter for the target option. If the value associated with a target probability is "relatively bad," then a positive difference in probability that exceeds the threshold does not increment the counter for the target option and instead increments an alternative option. Simple rules for determining the sign of the probability are described by Stewart and Simpson (2008) and Noguchi and Stewart (2018). Markov chain methods have been developed to compute the predictions of the model (Noguchi & Stewart, 2018).

Poisson race (PR) model The Poisson race (PR) model builds on the earlier horse race modeling ideas of Marley and Colonious (1992) and Townsend and Ashby (1983). Otter, Allenby, and Zandt (2008) proposed that each choice option i is associated with an independent counter $P_i(t)$, which counts the events (evaluations) that favor option i at time t. Events favoring a particular option occur at times distributed according to a Poisson process. (The model is a continuous time model, but we will describe its discrete time analog.)

Define $W_i(t)$ as a stochastic weight for evaluating option i at time t, which equals 1 when a favorable evaluation occurs, and otherwise is zero. The probability that $W_i(t)=1$ at each time step, which is called the *Poisson rate*, is independent of past evaluations. The stochastic weights for different options are assumed to be statistically independent. The Poisson rate r_i for option i is assigned to each option depending on its utility u_i, computed from its attribute values.

Then the count for option i at time t accumulates according to

$$P_i(t+1)=P_i(t)+W_i(t+1).$$

An advanatage of this model is that it provides straightforward mathematical solutions for the multialternative choice and decision time distributions for self-terminating tasks and fixed-time tasks.

Attention Drift Diffusion (ADD) model The Attention-Drift Diffusion (ADD) model (Krajbich, Armel, & Rangel, 2010; also see Krajbich & Rangel, 2011, Krajbich et al., 2012) was developed as a modification of Ratcliff's drift diffusion model (Ratcliff et al., 2016, and described in chapter 6). The modification was designed to account for attention biases produced by attending to an option during the choice process (e.g., looking at it). The option to which the person is attending is measured during the choice process by displaying each option (e.g., a picture of a food item) in a separate physical location and recording eye movements among the locations. Unlike the MDFT and LCA models, which describe attribute attention switching, the ADD model describes alternative attention switching.

First, it is assumed that each option has an unbiased mean drift rate, denoted as d_i, which represents the utility of the option when the person is looking at it. However, if the person is not looking at the option, then the mean drift rate is biased downward to $\theta \cdot d_i, \cdot 0 \leq \theta \leq 1$.

Define $l_i(t)$ as an attention variable with value 0 if the person is looking at option i at time t; otherwise, it is 1. It is important to note that the looking indicator $l_i(t)$ is an exogenous variable, and it is not derived or predicted by the ADD model. It is only observed and then used to determine when to change the mean drift rate. Using the attention indicator, we can define the moment-to-moment changing bias as $\theta^{l_i(t+1)}$, so that $\theta^1 = \theta$ when $l_i(t+1) = 1$ (not looking) and $\theta^0 = 1$ when $l_i(t+1) = 0$ (looking).

The ADD model uses the following process to accumulate preference for each option:

$$P_i(t+1) = P_i(t) + \theta^{l_i(t+1)} d_i + \varepsilon(t+1).$$

Similar to the DBS model, the ADD model uses the next best stopping rule: stop as soon as the first-ranked (maximum) preference state exceeds the second-ranked preference state by a threshold P^*. Currently, this model must be simulated, even for binary choice, because of the changes in drift.[1]

Associative Accumulator (AA) model Like MDFT and LCA, the Associative Accumulation (AA) model (Bhatia, 2013) assumes that one attribute is attended to for each time step, but the probability of attending to an attribute is not fixed across choice sets (as it is in MDFT and LCA). Instead, it depends on the activation strength produced by the association of attribute j with all the options presented in the choice set.

The activation strength of the jth attribute is defined as

$$c_j = a_0 + \sum_i \alpha_j \cdot v_{ij},$$

where the summation extends over all the options present and α_j is a scalar attached to attribute j.

As with MDFT, define the stochastic weight, $W_j(t)$, as the attention weight given to attribute j at time t, which is 1 or zero at each time step. Then the probability of attending to attribute j equals

$$p(W_j = 1) = w_j = \frac{c_j}{\sum_j c_j}.$$

Define $u_{ij}(t)$ as the utility of the value of option i on attribute j. Then the preference for option i is updated according to

$$P_i(t+1) = (1-\gamma) \cdot P_j + \sum_j W_j(t+1) \cdot u_{ij} + \varepsilon(t+1).$$

Like MDFT, but unlike LCA, there is no truncation, and the preference state $P_i(t+1)$ is permitted to go below zero. Also, unlike MDFT but like LCA, the preference states for binary choice do not sum to zero, so the dimension remains the same as the number of options. Again, this makes the model more complex, and even binary choice probabilities and decision times must be simulated in a two-dimensional plane.

Multialternative Linear Ballistic Accumulator (MLBA) model The Multiattribute Linear Ballistic Accumulator model (Trueblood, Brown, & Heathcote, 2014) is very different from the three previous stochastic accumulation models. This model is an extension of the Linear Ballistic Model (LBA) by Brown and Heathcote (2008), discussed in chapter 6. Recall that according to this model, each action corresponds to an accumulator, which races in parallel, and the one that reaches threshold P^* first is chosen. However, once a starting position of an action, $P_i(0)$, which is the intercept of a line at time zero, and the what they call the drift rate of an action d_i, which is equal to the slope of a line, are selected for each action, the accumulators deterministically integrate the slope over time $P_i(t) = P_i(0) + d_i \cdot t$. Thus, the time to reach the threshold for each action is simply the distance over rate $[T_i = (P^* - P(0))/d_i]$. The action with the shortest time is chosen for certain (actually, it is determined before the integration even starts). To make the model probabilistic, it is assumed that the starting point and the drift rate varies across the trials according to some specified distributions.

The contribution of MLBA is to include the assumptions needed to account for context effects. Most of the contextual processing that occurs in the MLBA model occurs at the front end during the coding of the mean of the drift rates, d_i. The rest is the same as the standard LBA model.

The MLBA model is described here for two attributes, $j = 1, 2$. The first step in the model is to define the utility of an alternative on an attribute. The utility of alternative i on attribute j is symbolized as u_{ij}, which is determined by a nonlinear map, which involves a single curvature parameter that can vary the convexity/concavity of the utility function.

The second step is to determine the model weights. A weight is assigned to each attribute j for each pair of alternatives (i,k) as follows:

$$w_{j,(i,k)} = \exp\,[-\lambda_j \cdot |\,[u_i - u_k\,|],$$

where $\lambda_j = \lambda_1$ if $u_{ij} - u_{kj} > 0$, and $\lambda_j = \lambda_2$ if $u_{ij} - u_{kj} < 0$.

Third, a comparison value, c_{ik}, is computed for each pair of alternatives (i, k), i not equal to k:

$$c_{(ik)} = w_{1,(i,k)} \cdot (u_{i1} - u_{k1}) + w_{2,(i,k)} \cdot (u_{i2} - u_{k2}).$$

Fourth, what is called the mean drift rate d_i is assigned to each action:

$$d_i = I_0 + \sum_{i \neq k} c_{(ik)}.$$

The mean drift rate, d_i, determines the mean of a normal distribution of drift rates that vary across trials. At this point, the standard LBA model does the rest of the work. Currently, the MLBA model is restricted to two attributes. An advantage of the ballistic nature of MLBA is that simple mathematical formulas are available for computing multiple alternative choice and decision times.

Several other models have appeared (Bergne et al., 2019; Tsetsos, Chater, & Usher, 2012; Solway & Botvinick, 2015; Diederich & Trueblood, 2018). However, these models were newly developed or developed for only limited and specialized applications at this point.

8.2 Comparison of Models

In this section, we first compare these models on their accounts of findings concerning context effects from multialternative choice (see also Wedell, 2015). After that, we consider their accounts of findings from risky choices, findings that relate choice and decision time, and findings that relate choice and eye movement measurements. We can focus on only some of the major findings; there are many variations and moderations of the phenomena that these models have addressed that are beyond the scope of this book.

Figure 8.1 illustrates the basic designs for investigating context effects. The key idea is to examine the effect of adding a new option to an existing set of choice options. This graph assumes that choice options are described by two major attributes (e.g., attribute 1 could be the economic attributes of a consumer product, and attribute 2 could be the quality of an economic product). Options *A* and *B* are the targets for context effects. Adding option *S* to the set {*A*, *B*} placed close to *B*, produces a similarity effect by increasing the probability to choose option *A*; adding option *C* to the set {*A*, *B*} makes *A* seem like a compromise and increases the probability to choose option *A*; adding option *D* to the set {*A*, *B*}, which is defective with respect to *A*, increases the probability to choose option *A* to produce an attraction effect.

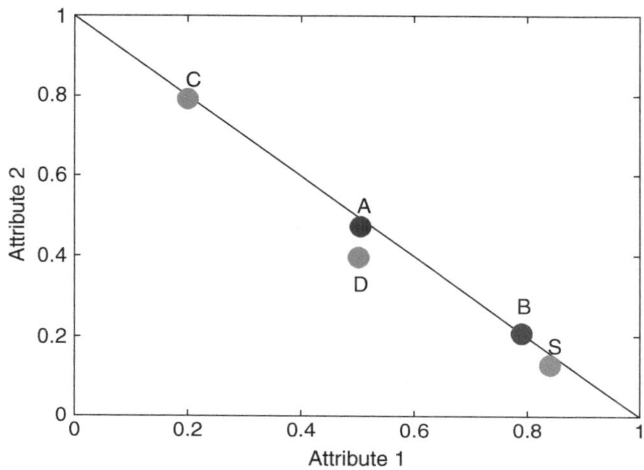

Figure 8.1

Illustration of choice options used to investigate context effects. Options *A* and *B* are target options. Adding option *S* to the set {*A, B*} produces a similarity effect. Adding option *D* to the set {*A, B*} produces an attraction effect. Adding option *C* to the set {*A, B*} produces a compromise effect.

Similarity effect Recall from chapters 3 and 7 that the similarity effect refers to the fact that when presented with a choice among three options, *A*, *B*, and *S*, where *S* is competitive with and similar to *B*, then *S* lowers *B*'s probability of choice (Tversky, 1972). MDFT, LCA, and AA models provide sampling accounts for the similarity effect by the attention-switching mechanism. This mechanism induces a positive correlation between $(P_A - P_B)$ and $(P_A - P_S)$ differences, which produces an advantage for option A. In contrast, the MBLA accounts for the similarity effect by the combination of weights and comparison values that is formed before deterministic and dynamic process even begins. The DBS model accounts for this effect by assuming that small differences between values fail to exceed a threshold in order to produce a winning comparison value. The PR and ADD models have never been applied to this effect to date.

Attraction effect Recall from chapters 3 and 7 that the attraction effect refers to the fact that when presented with a choice among three options, *A*, *B*, and *D*, where *D* is similar to but defective with respect to *A*, then *D* bolsters *A* and increases its probability of being chosen (Huber, Payne, & Puto, 1982). MDFT provides dynamic processing to account for this effect:

the negative valence from the defective option feeds back though the negative lateral inhibitory links to produce a bolstering effect. The LCA model accounts for this more directly by the loss aversion utility function used to compare options. The AA and MLBA models account for this immediately before processing: the AA model uses the associations to produce weights that support the stronger attribute for the dominating option *A*; the MLBA uses the weights, $w_{j(ik)}$, that are formed right away in order to put more weight on the stronger attribute of the dominating option. The DBS model accounts for this by the changing the probability of a winning comparison for the dominant option when the decoy is introduced. The PR and ADD models have never been applied to this effect either.

Compromise effect Recall from chapter 7 that the compromise effect refers to the fact that when presented with a choice among three options *A*, *B*, and *C*, where *B* is extreme on one attribute, *C* is extreme on another attribute, and *A* lies between the extremes on both attributes, then the compromise option *A* has an increased probability of being chosen (Simonson, 1989). MDFT uses lateral inhibitory feedback processing among the compromise and the extremes to induce a positive correlation between $(P_A - P_B)$ and $(P_A - P_C)$ differences, which produces an advantage for option *A*. The LCA model once again uses loss aversion to account for this effect. The AA model again uses the associations to produce weights that support the stronger attribute for the compromise option, *C*. The MLBA model accounts for the compromise effect by the utility function that maps values into utilities. The DBS model accounts for this via changes in the probability of sampling a winning value on an attribute for the compromise option when an extreme option is introduced. Once again, the PR and ADD have never been applied to this effect.

Reference point effect Recall from chapter 3 that the reference point effect refers to the fact that when given a choice among options *A*, *B*, and R_A, where R_A is similar to but less attractive than *A*, then *A* is chosen more frequently than *B*; but when given a choice among options *A*, *B*, and R_B, where R_B is similar to but less attractive than *B*, then *B* is chosen more frequently than A (Tversky & Kahneman, 1991). MDFT, LCA, AA, DBS, and MLBA account for this effect using the same mechanisms used to account for the attraction effect. As with the rest of the context effects, the PR and ADD models have never been applied to this effect.

Context effects under time pressure The compromise and attraction effects have been found to decrease with increasing time pressure using consumer-choice types of tasks (Pettibone, 2012). The same holds for reference point effects (Ashby et al., 2012). MDFT, LCA, DBS, AA, and MLBA models can account for this effect. According to MDFT, the lateral inhibitory process takes time to build in order to produce these effects. According to LCA and AA, the larger input produced by loss aversion (LCA) or association (AA) for the dominating or compromise option gets integrated over time. DBS accounts for this by reducing the threshold for stopping under time pressure, which leads to more random choices. According to MLBA, lengthening the time of the linear ballistic process increases the advantage provided to a larger mean drift rate for the attraction and compromise options. The similarity effect also has been found to increase with deliberation time but using an inference type of choice task (e.g., inferring guilt or innocence, Trueblood et al., 2014). MDFT, LCA, and AA do not predict this effect, but MLBA and DBS do.

Affective versus nonaffective choices The studies by Trueblood (2012) and Trueblood and colleagues (2013) show that the three standard context effects from the consumer choice literature also occur in inference and perceptual choice tasks. The existence of all three context effects in inference and perception calls into question the loss aversion explanation of LCA. In these experiments, the attributes are nonhedonic, and there is no notion of gain or loss in which loss aversion could operate. One could possibly redefine LCA's loss aversion as an asymmetric weighing of advantages and disadvantages on a particular dimension, but this strays far from the original idea.

Compromise and correlation Recall that according to MDFT, the compromise effect is explained by a correlation between extreme options that arises during processing from lateral inhibition, which results in an advantage for the compromise. Usher, Elhalal, and McClelland (2008) experimentally tested this by giving participants a trinary choice from a compromise type of choice set, but after the selection was made, they told participants that the chosen option was no longer available. Using this design, they argued that if extreme options are correlated, as in MDFT, then when one extreme is initially selected, the other also has a high level of activation; thus, it is more likely to be selected than the second choice. However, experimental results showed that the compromise option was still selected more often.

Hotaling, Busemeyer, and Li (2010) argued that MDFT can account for these results because of individual differences in asymmetric attention weighting. If a person initially chooses a first extreme option which is extreme on the first dimension, then that person is not paying much attention to the second attribute, which is extreme for the second extreme option, and so the second extreme option has low valence when the first extreme option is initially chosen. This is why the compromise option is chosen over the second extreme option when the first extreme option is eliminated.

Choice deferral effects The term *choice deferral* refers to postponing a decision to a later time or avoiding the decision altogether. This is a common feature of everyday decision-making, which has been extensively investigated (see Bhatia & Mullet, 2016, for a recent review). Research on this problem has discovered a number of moderating effects when choice deferral is included as an option. For example, the direction of context effects can change depending on whether choice deferral is available. A few attempts have been made to model these moderating effects using sequential sampling models. In particular, Jessup and colleagues (2009) proposed adding a stopping time limit to DFT to account for some of these effects. However, Bhatia and Mullet (2016) have demonstrated that combining the associative accumulating mechanism of Bhatia's (2013) AA model provides a broader and more systematic account of all the known findings. Additionally, Bhatia and Mullet (2016) quantitatively fit a time limit-based accumulation model to the experimental data using the associative accumulation mechanism, and they found that their model was able to provide an accurate account of choice and deferral probabilities and decision times.

Value context effects The term *value context effects* refers to changes in subjective values (utilities or subjective decision weights) produced by manipulating the distribution of objective values experienced by the decision-maker (Stewart et al., 2003; Rigoli, 2019). For example, changing the recently experienced probability and outcome values can reverse later preferences between gambles (Stewart, 2009). DBS provides a front end for determining subjective values that go into the sequential sampling process. Its main contribution is to develop a sampling theory that describes how utility functions, weighting functions, and discounting function arise from this sampling procedure (Stewart, Reimers, & Harris, 2014) and how these functions can change with context (here, the word *context* refers to the distribution of experienced values). Rigoli (2019) has proposed a new, and perhaps more comprehensive, model for these effects. None of the other models have addressed these kinds of context effects directly.

Risky decision paradoxes In chapter 3, a number of risky decision-making paradoxes, including the Allais paradox, were summarized. Both DFT (Bhatia, 2014; Johnson & Busemeyer, 2016) and DBS (Stewart & Simpson, 2008) provide accounts for a wide range of findings. However, only DFT has provided an explanation for variance effects (Busemeyer, 1985) and correlation effects (Diederich & Busemeyer, 1999; Andraszewicz, Rieskamp, & Scheibehenne, 2015) on choice under uncertainty. None of the other models have been applied to risky decisions or choice under uncertainty at this point.

Choice-decision time relation Although all the models are capable of making predictions for both choice and decision time, only a few of them have actually been used to predict both choice and decision time data. DFT has been used to fit binary choice probabilities accurately, and then it used the same parameters to predict mean response times in several applications (Busemeyer & Townsend, 1993; Dror, Busemeyer, & Basola, 1999; Diederich & Busemeyer, 2003); in addition, it has been used to simultaneously fit choice and decision time distributions (Dai & Busemeyer, 2014). The PR model was successfully used to simultaneously fit choice and response time distributions for consumer-choice tasks (Otter, Allenby, & Van Zandt, 2008). The ADD model has been used to fit choice and response times accurately for both binary and triadic choices between food products (Krajbich, Armel, & Rangel, 2010; Krajbich et al., 2012; Milosavljevic et al., 2010) and consumer products (Krajbich et al., 2012), as well as social decisions such as the dictator game (Krajbich et al., 2015). Recently, the MLBA model has been used to fit choice and decision time distributions for a wide range of experiments (Evans, Holmes, & Trueblood, 2019; Molloy et al. 2019). The original version of the LCA has been fit to both choice and response time data for perceptual choice, but the LCA has not yet been applied to either choice or decision time data. Neither the DBS or AA model has been applied to decision time at this point.

Eye movements There are two methods of analyzing eye movements that can be used to examine dynamic preference models with multiattribute, multialternative choice. One method is to present a set of options and record the sequence of eye movements to attribute values that describe the options, and finally to analyze the pattern of movements across alternatives and attributes (Payne, Bettman, & Johnson, 1993). Presumably, this gives an idea of the sequence of attention to information about the attributes and their values on each alternative in the choice set. Another method is to examine how changes in eye movements from one option to another affect the final choice that is made.

Using the first eye movement method of analysis, a critical processing assumption of MDFT, LCA, and MLBA is that comparisons across the options are made within an attribute. The PR, AA, and ADD models predict the opposite—processing across attributes within an option to produce a single drift rate each option. In fact, comparison across options within an attribute is found to be the dominant eye movement pattern. Noguchi and Stewart (2014) reported that participants first made pairwise comparisons of options within one attribute and then moved to the next attribute. The comparison across attributes accords well with MDFT, LCA, DBS, and MLBA, but not with AA and ADD. MDFT and LCA also assert that once an attribute is selected, all the options are compared rather than just a pair of options. But Noguchi and Stewart (2014) found that only pairs are considered at one time. It may not be difficult to relax this assumption and allow sampling of pairs of options about an attribute. In particular, for MDFT, if one replaces the original valence assumption (which computes a valence vector for all options in parallel for a single time step) with the expectation of valence produced by randomly sampling one of the pairs, then the results are practically equal.

Using the second eye movement method of analysis, the ADD model has an advantage because it predicts several quantitative empirical relations between choice and looking time that are empirically observed. For example, the ADD model accurately predicts the observed finding of a strong and increasing relation between excess time looking at an option and the choice that is made, and it predicts the observed increase in probability of choice, given that the chosen option was the last to be viewed. The findings clearly demonstrate strong relations between eye movements and choice. Smith and Krajbich (2018) also tested the consistency of the eye movement relationship to choice across different domains (e.g., money or food) in the same person. Individuals with a strong relation in one task were likely to have strong relations in the others. Only the DBS model and the ADD model have attempted to model the connections with eye movements.

Individual differences One issue that that needs to be addressed is whether context effects actually occur at the individual level, and if they do, whether there are large individual differences between these effects (Hutchinson, Kamakura, & Lynch, 2000). A within-participant experiment by Soltani, De Martino, and Camerer (2012) reported systematic context effects (similarity and attraction) at the individual level, and so context effects are not just aggregation artifacts. However, large individual differences in the types of context effects have also been found (Liew, Howe, & Little, 2016). For example, the similarity effect tends to be negatively correlated with the compromise effects (which was actually predicted by MDFT). Therefore, it is important to try to analyze these effects on the individual level or include some type of classification of individual differences.

Quantitative model comparisons So far, no one has quantitatively compared all these models, but various types of quantitative model comparisons have been made. In a couple of studies, the models were fit to individual data. Rieskamp (2008) found that DFT predicts choice probabilities for binary choices between gambles better than a probabilistic version of prospect theory (using a logistic choice function) and a stochastic version of the priority heuristic model. Scheibehenne, Rieskamp, and Gonzalez-Vallejo (2009) reported that DFT predicts binary choices between gambles better than the proportional difference model (Gonzalez-Vallejo, 2002).

Other studies were fit to aggregate data. Berkowitsch, Scheibehenne, and Rieskamp (2014) found that MDFT predicts trinary choices better than a multivariate probit or multinomial logit model when context effects were present. Hancock, Hess, and Choudhury (2018) determined that MDFT predicts better than a multinomial logit model using stated preferences for travel route choices.

Krajbich and colleagues (2010) and Krajbich and Rangel (2012) found that the ADD model predicts choice and response times (when including changes in drift with changes looking) better than a standard diffusion model (Ratcliff, 1978). In these cases, the models were applied to individual data. Krajbich and colleagues (2015) concluded that the Weiner drift diffusion model fits better than the Ornstein-Uhlenbeck (OU) drift diffusion model.

Bhatia (2017) found that the AA model predicted binary choice probabilities better than a probabilistic version of prospect theory (using a logistic choice function) in two large experiments investigating reference point effects. In this case, the models were fit to individual data.

Trueblood and colleagues (2014) found that the MLBA fits trinary choices from inference and perceptual tasks better than MDFT. In this case, the models were fit to aggregate data. However, Hotaling and Rieskamp (2018) found that MDFT and MLBA produced equally accurate predictions for preferential choice on the individual level.

Turner and colleagues (2018) conducted a comprehensive hierarchical Bayesian comparison of models using a preferential choice task and found that MDFT predicts aggregate data best, but the AA and MLCA perform better than MDFT and MLBA when using a hierarchical Bayesian analysis of individual differences. They also examined a combination of model assumptions and found that mechanisms such as lateral inhibition, loss aversion, and pairwise attribute comparisons are some of the most important principles to include in models.

Most recently, Evans and colleagues (2019) conducted a comprehensive comparison of the MDFT, MLCA, AA, and MLBA models in twelve studies of context effects. They said that the MDFT and MLBA models fit equally well, and both of these models fit better than the MLCA and AA models when applied only to data from choice probabilities. However, when both choice and response time distributions were analyzed simultaneously, greater differences among models were found. For the attraction effect, all the models predicted equally well. For the similarity and compromise effects, all the models did well at predicting the response time distributions, but only MLBA simultaneously provided accurate predictions for the choice proportions. One reason that the MLBA has an advantage in modeling choice and response time distributions is because it has between-trial variability in the drift rate, which was not included with the other competing models.

Table 8.1 summarizes these model comparisons. As can be seen, the results are not very consistent. In sum, it is too early to draw conclusions from these initial findings and much more quantitative model comparison needs to be done.

A different type of comparison was carried out by Colas (2017). Several neurally plausible models were compared with more traditional cognitive-based models based on choices between foods. The neurally plausible models

Table 8.1
Summary of model comparisons.

	Aggregate	Individual	Hierarchical
Rieskamp (2008)		DFT > PT, PH	
Scheibehenne et al. (2009)		DFT > PD	
Berkowitsch et al. (2014)	MDFT > MNL, probit		
Hancock et al., 2018	MDFT > MNL		
Bhatia (2014)		AA > PT	
Trueblood et al. (2014)	MLBA > MDFT		
Hotaling et al. (2018		MDFT = MLBA	
Turner et al. (2018)	MDFT>AA>MLCA>MLBA		AA, MLCA > MDFT, MLBA
Krajbich et al. (2010, 2012, 2015)	ADD > DDM, OU		
Noguchi and Stewart (2018)			MDFT = MLBA=MDBA
Evans et al. (2019)			MLBA > MDFT, MLCA. AA

PT = prospect theory, PH = priority heuristic, PD = proportional difference model, AA = associative accumulator model, MDFT = multiattribute decision field, MNL = Multinomial logit model, MLBA = multialternative linear ballistic model, MLCA = multialternative leaky competing accumulator model, MDBS = multialternative decision by sampling, AAD = attention drift diffusion model, DDM = drift diffusion model, OU = Ornstein—Uhlenbeck model.

were found to be superior to more traditional race and drift-diffusion models, both quantitatively and qualitatively, despite remaining parsimonious.

Summary MDFT, DBS, LCA, AA, and MLBA provide accounts of the major context effects, but the PR and ADD models have never addressed these effects. It is not clear whether the PR or ADD models can account for any of the context effects in their current form for the following reason: They are built on the assumption that a drift rate is assigned to each option in a context-free manner. In other words, the drift rate for option *A* when presented a choice between options A and B remains the same as the drift rate for option A when there is a choice between A, B, and C.

This is analogous to the context-free value assumption that is built into the Luce choice model discussed in chapter 3, which led its empirical problems. Only DFT and DBS have been applied to risky choice paradoxes. The DBS model has an advantage concerning the value-related context effects on utilities and decision weights. Noguchi and Stewart (2018) argued that the MDBS model accounts for the largest variety of context effects among the models reviewed here. Only DFT and DBS have been applied to risky decisions and decisions under uncertainty. ADD has been applied to a number of social decisions, including the dictator game. DFT, LBA, PR, and ADD have been applied simultaneously to predicting choice and decision time results for preference (the original LCA was applied to perceptual choice, but not the newer LCA with loss aversion). A unique advantage of ADD and DBS is their ability to account for the effect of eye movements on choice.

9 Experience-Based Choice

Many choices that people make are based on their experience with the outcomes of same or similar past choices. For example, deciding whether to use an online web service depends on past experiences with similar services; deciding how much food to eat to avoid gaining weight depends on past eating habits; deciding whether to speed on a highway depends on past experience with the law enforcement on that highway; and deciding how to deal with an upset child depends on past experience with different coping strategies.

This chapter describes models that integrate learning into the choice process (see Dayan & Abbot, 2001, for a comprehensive review). Models of experience-based decision-making were quite popular in the 1950s and 1960s: Estes (1950) introduced stimulus-sampling learning models, and Bush and Mosteller (1955) introduced linear operator learning models. However, interest in learning and decision-making waned for a period of time. Research in learning moved in new directions using connection/neural networks to model categorization, word learning, and other cognitive activities (Rumelhart, Hinton, & McClelland, 1986; Grossberg, 1988). Research in decision-making refocused on description-based choice tasks (see chapter 2), providing verbal and textual descriptions that attempted to eliminate the need for learning. Models of experience-based decision-making were eventually picked up again in the 1990s (Busemeyer & Myung, 1992; Erev & Roth, 1998). Two findings particularly stimulated research on experience-based decision-making: one concerns preference reversals between choices based on description versus experience (the description-experience gap); the other concerns the relation between learning models for decision-making and neuroscience (the error signal–dopamine link). We explore the description–experience gap in this chapter, and we will discuss the error signal–dopamine link in the next chapter.

Models of experience-based decision-making usually investigate decision-making under uncertainty, in which each action produces one of several possible outcomes but the probability of the outcomes are initially unknown. First, we describe research on repeated decision tasks, also known as *multiarm bandit problems* (Berry & Fristedt, 1985). These are situations involving a single choice followed by a payoff, and the same choice problem is repeated for several hundred trials. Trial-by-trial feedback provides the opportunity to learn the payoff distributions associated with each action. Second, we review some issues that arise when experience-based choices are compared to choices based on description. Preference reversals can occur between experience and description, even when the outcomes seem probabilistically identical. Finally, we review complexities that arise with planning during multistage or dynamic decisions, also known as *Markov decision problems (MDPs)*. There is a huge body of literature on experience-based decision-making, so we will only present the big picture, and we won't be able to delve deep into many of the complex findings in this field.

9.1 Multiarm Bandit Problems

One way to experimentally study experience-based decisions is to use the multiarm bandit problem. In a typical experiment, in each trial, the decision-makers are asked to choose from n alternatives with initially unknown outcome probabilities, and then they make a choice and experience the payoff for that choice. The same task is repeated with the same choice set for several hundred trials. Initially the choices are random across options, but with experience, each learner eventually forms a preference manifested by the probability of choosing each option across trials. For a review of the research findings with this task, see Erev and Roth (2014).

A very popular example is the Iowa gambling task (IGT), developed by Bechara, Damasio, Damasio, and Anderson (1994), which is described next. The task was designed to test Damasio's (2006) hypothesis regarding the role of emotional feelings that arise early in training on the IGT task, which are presumably processed in the orbital frontal cortex. For each trial, the undergraduate student participants were asked to choose a card from one of four decks. The cards in the deck described the payoff received if that card is chosen, but the contents of the decks were initially unknown. After each participant picked a card and turned it over, the payoff for that choice was revealed. The payoffs for the four decks, labeled *A, B, C,* and *D,* were as follows:

A: Win \$1 on each trial, but lose \$2.50 lose on half the trials

B: Win \$10 on each trial, but lose \$12.50 lose on one out of ten trials

C: Win \$0.50 on each trial, but lose \$0.50 on half the trials

D: Win \$0.50 on each trial, but lose \$2.50 on one out of ten trials

Each expected payoff for decks *A* and *B* is –\$0.25, and each expected payoff for decks *C* and *D* is \$0.25. This task is unusual because the person wins something and loses something on each trial (typically, experiments only show a single win or a single loss and not both). The separate display of the win and loss has an effect on the way people process the gains and losses. This problem is a bit tricky: normal players initially focus on the win and initially choose a card from one of the bad (negative expected value) decks *A* or *B* more frequently, but after many trials, they learn the penalty and switch to choosing cards from the good (positive expected value) decks *C* and *D* more frequently. An important finding is that patients with damage to the orbital cortex of the brain, as well as cocaine abusers, never learn to switch from bad decks, *A* and *B*, to the good decks, *C* and *D* (Bechara et al., 1994; Grant, Contoreggi, & London, 2000). Figure 9.1 shows the proportion of choices from decks *C* and *D* from Grant et al. (2000). These findings have stimulated many applications of this task for research on neuroscience and drug abuse decision-making.

Early-stage models The models initially developed for this task assumed that each action *i* was associated with an expectation or preference estimate, P_i, for an action. The preference estimates for each action were learned and used to generate choices according to four basic principles. (See the appendix for an example program.)

The first was a subjective evaluation of the payoff experienced after a choice is made. For example, as described in chapter 2, a prospect utility function could be used with risk aversion (represented by coefficient α) and loss aversion (represented by coefficient λ):

$$u(x) = x^{\alpha} \text{ for } x > 0\text{ ; } u(x) = -\lambda \cdot |x|^{\alpha} \text{ for } x < 0, \tag{1}$$

where x is the net payoff experienced after a trial. The parameters are usually constrained to $0 < \alpha < 1$ and $0 < \lambda < 5$.

Second, an updating rule was used to update the preference estimate (denoted as Q_i for action *i*) based on the experienced utility. Define $\delta_i(t) = 1$ if action *i* is chosen on trial *t*, and zero otherwise. Two updating rules have been used frequently. One popular learning rule (used, for example, by

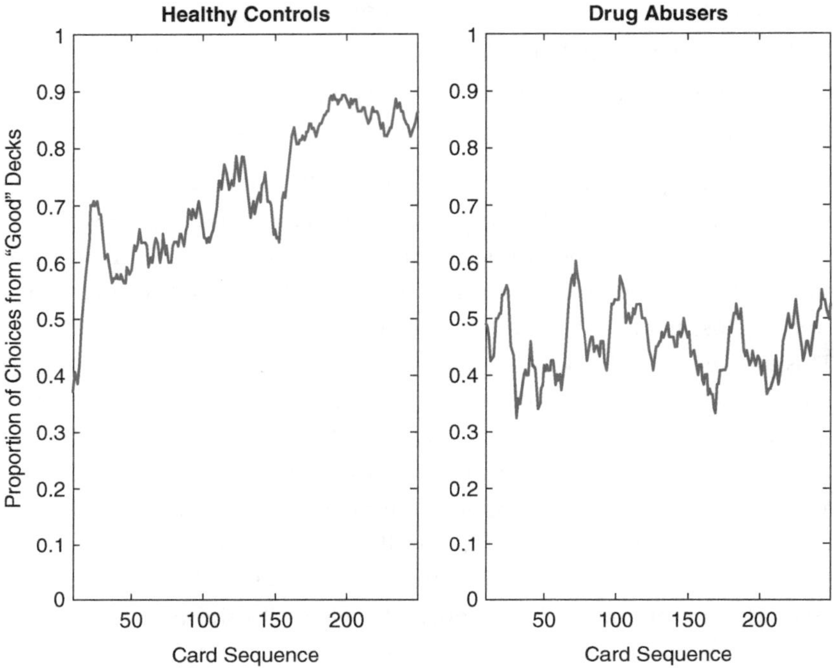

Figure 9.1
Proportion of choices from the good decks (decks *C* or *D*) with positive expected value on the IGT task across trials for nonabusers (left panel) and cocaine abusers (right panel).

Busemeyer & Stout, 1992, for the IGT task) is called the *delta learning rule*, which updates the preference only for the action that was chosen on trial *t* according to the following rule:

$$Q_j(t+1) = Q_j(t) + \beta \cdot [u(t) - Q_j(t)] \cdot \delta_i(t). \qquad (2)$$

The term $[u(t) - Q_j(t)]$ is the error signal that increases preference if the experienced utility is greater than the current preference estimate, or decreases preference if the experienced utility is less than the current preference estimate. Another popular rule is the *decay reinforcement rule* (Erev & Roth, 1998), which applies to all actions on every trial:

$$Q_j(t+1) = \beta \cdot Q_j(t) + u(t) \cdot \delta_i(t) \ . \qquad (3)$$

The learning coefficient, β, is usually constrained to $0 < \beta < 1$ for both models.

Third, an autoregressive or persistence term can be averaged with the current preference state $Q_i(t)$ to produce a bias to repeat the previous choice (Worthy, Pang, & Byrne, 2013). The persistence term can be written as

$$R_i(t+1) = \theta \cdot R_i(t) + \eta \cdot \delta_i(t), \tag{4}$$

with $0 < \theta < 1$ and $-1 < \eta < 1$. This persistence term is combined with the current preference state to produce a biased preference state:

$$P_i(t) = w \cdot Q_i(t) + (1 - w) \cdot R_i(t), \tag{5}$$

where $0 < w < 1$.

Fourth, a choice model is used to map the biased preference estimates for the actions into choice probabilities. Usually, the Luce ratio of strength model, discussed in chapter 3, is used:

$$p(\text{Choose } i \text{ on trial } t) = \frac{e^{\gamma(t)\cdot P_i(t)}}{\sum_j e^{\gamma(t)\cdot P_j(t)}}. \tag{6}$$

Parameter $\gamma(t)$ controls the degree of exploration/exploitation or the consistency of the mapping from preference to choice probability. In some models, it is assumed that $\gamma(t) = \gamma$ is constant across trials, and in others, it is assumed to change across trials according to $\gamma(t) = \gamma \cdot t^c$, with $-2 < c < 2$. Increasing $\gamma(t)$ across trials changes the choices from exploration to exploitation; decreasing $\gamma(t)$ produces the opposite effect (e.g., if the person is getting bored and starts responding randomly).

Model applications One purpose for using a learning and choice model is to provide a mechanistic explanation for the decision-making deficits exhibited by brain-damaged and drug-abusing populations on experience-based choice tasks like the IGT. The parameters of the learning models described here can be estimated for individuals using maximum likelihood from the trial-by-trial choices. More recently, hierarchical Bayesian models have been applied (Ahn et al., 2011). These individual parameters then provide a decomposition of the performance on experience-based decision tasks like the IGT task into utility parameters such as loss aversion, learning parameters such as the learning rate β, and choice parameters such as the exploration-exploitation parameter γ.

In one study (Yechiam et al., 2005), these three kinds of parameters were used to characterize differences in performance on the IGT between special populations that included patients with ventral medial prefrontal cortex (VMPC) damage, cocaine abusers, Huntington disease patients, Parkinson disease patients, young poly (multiple types of drugs) drug users, elderly, and Asperger syndrome individuals, and control groups. Although several of these populations produce the same type of poor performance on the IGT, the parameters reflected different possible mechanisms. For example, the VMPC patients produced differences in the learning parameter, but the cocaine abusers produced differences in attention to losses.

The learning and choice models are also useful for performing model-based functional magnetic resonance imaging (fMRI) analyses (e.g., O'Doherty et al., 2003). For example, the parameters of a learning model can first be estimated from the trial-by-trial choices on an experience-based choice task. Then these parameters can be inserted into the model to generate a time series of predictions that provide a predictor variable for a linear time-series model, which can be used to predict blood-oxygenation-level-dependent (BOLD) signals across time in brain regions of interest. For example, Ahn et al. (2011) used hierarchical Bayesian estimation methods to estimate model parameters with thirty participants. Then the individual parameter estimates were used to generate predicted choice probabilities on each trial from the learning model for each person. The predicted choice probabilities were convolved with a canonical hemodynamic response filter to produce the model predictions. The time series from the model was then used as a predictor in a linear time series model, along with other behavioral measures or task outcomes, to identify and predict fMRI BOLD signals. It turned out that the model predictor was a better predictor than several types of behavioral measurements or outcomes from the IGT task.

Model comparisons Several studies have conducted comparisons of models produced by various combinations of these four principles using data from the popular IGT task. Yechiam and Busemeyer (2008) compared models separately for each of 162 participants using Bayesian information criterion (BIC) statistics. The BIC was obtained from the maximizing the log likelihood of the trial-by-trial predictions for the next choice based on the past history of payoffs. The BIC comparison found that the decay reinforcement learning rule [equation (3)] performed better than the delta learning rule [equation (2)]; however, this model comparison did not include the persistence factor [equation (4)]. Steingrover, Wetzels, and Wagenmakers (2016) compared models separately for each of 771 participants using Bayes factors (estimated using importance sampling methods rather than asymptotic BIC estimates). The Bayes factor comparison found that the overall evidence favored a model using the delta learning rule [equation (2)] when it was combined with a persistence term [equation (5)].

A different method for comparing models is to use the generalization criterion (Busemeyer & Wang, 2000). This method is similar to a cross-validation method (Browne, 2000), but it differs in one important aspect. Cross-validation divides all the design points randomly into two subsets: calibration and cross-validation; the parameters are estimated from the calibration subset, and these same parameters are used to make predictions for the cross-validation subset. Using cross-validation, each design point has an equal chance of being assigned to either the calibration or the cross-validation subset. This allows many cross-validation points to lie close to and between calibration points, making it easy for a model to interpolate from calibration to cross-validation. Generalization also divides the design points into two subsets, but not randomly: instead, the generalization points are systematically chosen to be outside a region of calibration points such that the model must extrapolate far beyond any of these calibration points.

The learning models described thus far have been compared using the generalization criterion. Ahn and colleagues (2008) performed generalization tests by applying the models to two learning tasks: one was the IGT described earlier, and the second was the Suzhou gambling task (SGT). Like the IGT, the SGT has four decks; two have the same positive expected values as the IGT, and the other two have the same negative expected values as the IGT. But unlike the IGT task, the Suzhou task presents only one net payoff and the payoff amounts and probabilities are different. The choice results produced by the SGT turn out to be quite different from those produced by the IGT. However, if a learning and choice model is capturing basic learning mechanisms that a person uses across tasks, then the same parameters should be applicable to both tasks. Ahn et al. (2008) compared models separately for each of thirty-six participants using log likelihoods of the trial-by-trial one-step-ahead predictions. The parameters were estimated from one task and then used to predict the other. The results turned out to favor the decay reinforcement learning model [equation (3)] over the delta learning model [equation (2)]. However, this comparison did not include the persistence factor.

Yechiam and Busemeyer (2008) performed model comparisons using the generalization method with three binary choice problems: a choice between two actions with equal variance but different expected values; a second choice between a risky action (i.e., a small probability to win a big payoff) versus a sure thing; and a third choice between a small expected value and small variance action versus a large expected value and large variance action. The parameters were estimated from one of the choice problems and then used to make predictions in the generalization phase to the other two choice problems. The results of the model comparison revealed that (1) generalization from any model was better than a simple baseline model only when the parameters were estimated from the third binary choice task and generalized to the first two choice problems; and (2) when generalization was made from the first choice problem to the other two problems, the delta learning rule performed better than the decay reinforcement learning model. Once again, the persistence factor was not included in these comparisons.

In summary, the model comparisons result in different conclusions depending on the methods and tasks. This mixed conclusion indicates the need for both better methods *and* better models. The next section, on experience-based choice models, presents a more challenging method for model comparison and more sophisticated learning models.

Newer models and comparisons Erev, Ert, and Roth (2010) organized an international choice prediction competition. They conducted two experiments, and each included sixty randomly selected binary choice problems. The problems involved a choice between a certain medium outcome payoff M and a risky action that produced payoff H with probability p, and otherwise payoff L. An algorithm was used to generate a distribution of problems varying in the parameters (L, M, p), with one-third of the problems in the gain domain, one-third in the loss domain, and one-third a mixture of gains and losses. The values of M were selected from a distribution with a mean equal to the expected value of the risky action. The data set from the first experiment was posted on a website, which challenged other researchers to compete to predict the results of the second data set. Fourteen research teams responded to the challenge. The competitors proposed a model and used the first data set to calibrate their models. Then organizers used the models submitted by the contestants, as well as their parameters from the first data set, to generate predictions for the second data set. The winner for the experience-based choice problems was a model by Lebiere, Stewart, and West (2009). However, we present here a more recent extension of the winning model, called the Instance-Based Learning (IBL) model, by Gonzalez and Dutt (2011). The IBL model predicts even better than the winner did.

First, the IBL model uses the concept of persistence or inertia, which is a bias to repeat past decisions. In each repeated choice trial, the model selects either to repeat the last choice with probability p_I, or to choose the action with the highest accumulated preference $P_i(t)$, called the *blended value*. The blended value of action i is defined as

$$P_i(t) = \sum_j p_{ij} \cdot u(x_{ij}),$$

where p_{ij} is the probability of retrieving outcome x_{ij} from memory and $u(x)$ is a utility function. The probability of retrieval is defined as follows:

$$p_{ij}(t) = \frac{e^{A_{ij}}}{\sum_k e^{A_{ik}}}.$$

Here, A_{ij} is the activation of x_{ij}, which is defined as

$$A_{ij}(t) = \lambda \cdot \ln\left(\sum_{\tau \in T(x_{ij})} (t - \tau)^{-\delta} \right) + \varepsilon(t),$$

where the sum is over the set of time points in $T(x_{ij})$, which includes the time points at which the payoff x_{ij} occurred in the past with action i, and $\varepsilon(t)$ is a random noise term. The IBL model outperformed all the models submitted to the Erev et al. (2010) contest on repeated choice tasks. However, Plonsky, Teodorescu, and Erev (2015) showed that learners rely on experiences that followed the same *sequence* or *pattern* of outcomes preceding the current decision, rather than focusing only on a summary expectation from most recent experiences, as proposed by IBL.

Another advantage of the IBL model is that it has been applied not only to repeated decisions like the multiarm bandit problem, but also to another decision task called the *sampling paradigm*, which is described next.

9.2 Description-Experience Gap

The history of research on decision-making and choice has gone through several phases. Early in the 1950s and 1960s, researchers investigated these questions using the experience-based decision paradigm, but then, from the 1970s through the 1990s, decision researchers relied primarily on a description-based paradigm. This paradigm presents all the information required to make the decision in a verbal, textual, or graphical form so that no additional (experiential) learning is required. The decision paradoxes covered in chapter 2, such as the Allais paradox, are based on the description paradigm. However, starting around 2000, researchers became interested in comparing the results from the two paradigms because they seemed to disagree fairly dramatically. For example, consider a choice between risky action *A* (win \$4 with probability .80 or zero otherwise) and safe action *B* (win \$3 for sure). When presented descriptively (as in the previous sentence), most participants prefer the safe action; however, when learned from trial-by-trial feedback through repeated play, most participants prefer the risky alternative (Barron & Erev, 2003). This preference reversal between description and experience, for what seems to be the same formal gamble, is called the *description-experience gap* (Hertwig & Erev, 2009).

There are problems comparing results from the two paradigms because of the many differences that could contribute to the different outcomes. One problem is the obvious difference of repeated choice for the same problem used in many experience-based studies, as opposed to single choice for a problem used in many descriptive-based studies. It is well known that when offered a single choice to have a gamble played multiple times, people become less risk averse (Keren & Wagenaar, 1987). A way to counter this problem is to use an experience-based choice task that allows the decision-maker to learn the outcomes during an observational sampling stage and make a single final choice afterward. When comparing results from the sampling paradigm with the description paradigm, the discrepancy or gap remains (Hertwig et al., 2004).

Another problem is the possibility that the actual relative frequencies obtained from the experience-based paradigm can differ from the theoretical probabilities given in the description paradigm (Fox & Hadar, 2006). This is especially likely with small sample sizes and low theoretical probabilities. However, even when the experienced relative frequencies match the theoretical probabilities, the gap remains (Ungemach, Chater, & Stewart, 2009), as also occurs when participants are given large sample sizes from which to learn the probabilities (Hau et al., 2008).

A third possibility is that participants receive trial-by-trial feedback in the experience-based paradigm, but not in the description paradigm. To examine the role of feedback, Jessup, Bishara, and Busemeyer (2008) compared two conditions. In both, all the participants were provided full descriptive information and made the choices repeatedly; however, only one group received feedback after each trial. In this case, feedback was essential because there was a persistent reversal of preferences between the groups with and without feedback, keeping everything else constant.

The theoretical causes for the description-experience gap remain under debate, but the candidates include the following ideas (Hertwig & Erev, 2009): recency effects and the reliance on small, recently experienced samples; underestimation errors for rare experiences; reliance on similar past experiences (e.g., as proposed in the IBL model); and different decision strategies evoked by the different paradigms (Hau, Pleskac, & Hertwig, 2008).

9.3 Markov Decision Problems

The multiarm bandit problem described earlier is usually treated as a single-stage decision in which a choice is made, feedback is presented, and the trial ends. The initial learning models developed for this task relied only on the immediately obtained reward at the end of a trial. Dynamic decisions involve a sequence that has multiple decision stages, so later decisions depend on earlier decisions. But the simple reinforcement learning models discussed earlier in this chapter are not capable of making plans when a sequence of decisions is required. One important exception is the IBL model, which was originally developed for dynamic decision tasks (Gonzalez, Lerch, & Lebiere, 2003).

This section briefly reviews work on MDPs, which form a reasonably tractable type of discrete time and state dynamic decision task (Puterman, 2014). MDPs are characterized by a set of states S, a set of actions A, a set of payoffs X, a transition probability rule (T: $S \times A \rightarrow S$) mapping from state-action pair to a new state, and a reward probability rule (R: $S \times A \rightarrow X$) mapping from state-action pairs to payoffs. The Markov property comes from the assumption that the new state depends only on the previous state and action.

The decision problem has a finite number of stages N: at each stage of the problem, the decision-maker is in a known state and chooses a course of action, a reward/punishment is delivered, the state changes to a new known state, and the decision process continues until the end stage is reached. An example is a grid world (see figure 9.2), where the decision-maker starts at an initial point in the grid and attempts to take steps to reach a goal in another point in the grid world; the agent can move up or down or left or right, one step at a time, and each step in the grid can produce a positive or negative payoff according to a particular probability.

MDPs explicitly assume that the decision-maker knows the current state at each stage (e.g., they know their current location and the location of the goal in the grid world). On the other hand, if the current state is uncertain

.6 \$1, .4 − \$2	.5 \$1, .5 − \$1	.5 \$1, .5 − \$50	Goal \$100
.4 \$5,.6 − \$1	.5 \$10, .5 − \$10	.2 \$1, .8 − \$50	.5 \$1, .5 − \$1
Start	.5 \$1, .2 − \$50	.5 \$1, .5 − \$1	.5 \$1, .5 − \$1

Figure 9.2
Grid world example of an MDP problem. Each cell indicates the probability of winning or losing a specific amount of money when entering that cell.

(e.g., the person is not sure where they are located or where the goal is located in the grid world), then the problem is called a *partially observable Markov decision process (POMDP)*, and Bayesian inference is used to infer the current state. This is a more realistic but more complicated type of task.

Decision models for MDPs usually assume that the decision-maker's goal is to maximize the expected discounted utility accumulated across all N stages (see, e.g., Daw & Tobler, 2014). Suppose that the decision-maker uses a policy to choose actions for each state and define $X(t)$ as a random variable representing the payoff obtained on each stage for a given policy. Suppose that the decision-maker is located at state $s(t)$ and is considering action $a(t)$ at this stage. Then the decision-maker seeks the policy that maximizes the expectation:

$$Q(s(t), a(t) \mid policy) = E\left[\sum_{\tau=t}^{N} \lambda^{t-t} \cdot X(\tau) \mid s(t), a(t), policy\right],$$

where $0 < \lambda < 1$ is a discount factor (for short finite decisions, this can be set to 1.0). The optimal policy can then be expressed recursively as follows:

$$Q(s(t), a(t)) = E[X(t) \mid s(t), a(t)] \\ + \lambda \cdot \sum_{s} p(s(t+1) \mid s(t), a(t)) \cdot Q(s(t+1), a^{*}(t+1)), \tag{6}$$

where $a^{*}(t+1)$ is the action that is maximal for the new state $s(t+1)$. The quantity $Q(s(t),a(t))$ is called the *Q-value* for a state and action.

There are two general approaches for learning the policy that maximizes the expected discounted utility accumulated across all N stages (Dayan & Niv, 2008): one is called the *model-based method*, and other is called the *model-free reinforcement learning method*. First, consider the model-based method, which learns the entire structure of the MDP. The model-based method learns the state transition probability rule T and the reward probability rule R using Bayesian updating methods. Then the backward equation [equation (6)] is solved to find the Q-value for each state and action. The advantage of the model-based method is that it rapidly replans optimal paths when changes occur in the MDP. The disadvantage is that it requires solving equation (6) to determine the Q-values, which may be an unrealistic task for human decision-makers.

The model-free reinforcement method does not learn the structure of the MDP. Instead, it uses a local learning mechanism that gradually estimates the Q-value in a trial-by-trial manner based on an approximation to equation (6). Define $x(t)$ as the reward actually experienced by taking action $a(t)$ in state $s(t)$. Define $Q_t(s(t), a(t))$ as the estimated Q-value after t trials. The temporal-difference (TD) updating rule (Sutton & Barto, 1998) is used to update $Q_t(s(t), a(t))$ after taking action $a(t)$ in state $s(t)$:

$$Q_{t+1}(s(t), a(t)) = Q_t(s(t), a(t)) + \beta \cdot [x(t) + Q_t(s(t+1), a^*(t+1)) - Q_t(s(t), a(t))].$$

The term $[x(t) + Q_t(s(t+1), a^*(t+1)) - Q_t(s(t), a(t))]$ is the TD error signal, which differs from the earlier learning models for the multiarm bandit by including the forward-looking expected reward, $Q_t(s(t+1), a^*(t+1))$. The action $a^*(t+1)$ is usually set as being equal to the action that maximizes the estimated Q-value in the next state. The expected reward from the new state allows the learning process to occur so that the participant can learn to plan ahead. The temporal difference rule usually uses a Luce choice type (also called *soft max*) rule to probabilistically choose a response:

$$p(a(t) = a_j) = \frac{e^{\alpha \cdot Q_t(s(t), a_j)}}{\sum_k e^{\alpha \cdot Q_t(s(t), a_k)}}.$$

The temperature parameter α controls the degree of exploration versus exploitation. As mentioned earlier, this can increase across trials in order to change from an exploration to an exploitation regime.

A number of experiments have been conducted to compare model-free reinforcement learning with human performance in MDP-type problems, and the results provide some support for the TD types of learning mechanisms (see Walsh & Anderson, 2014, for a review). One interesting example is a pair of studies by Gureckis and Love (2009a, b). They challenged decision makers with options that appear attractive in the short-term but are non optimal in the long run by using short- and long-term rewards that were in conflict. Their work showed that the effectiveness of learning from reinforcement depends on the way participants represent the state of the MDP problem. They found that perceptual cues, which readily align with the underlying state of the task environment, helped participants learn to give up the short-term smaller rewards for more optimal long-term rewards. A TD learning model model achieved better fits than an elibility trace model based on decaying memory of past reinforcements, and the TD model closely matched the trial-by-trial dynamics of learning for the different cue conditions.

Model-based learning has also been shown to be operating in many planning under uncertainty tasks (Daw, Niv, & Dayan, 2005; Simon & Daw, 2011; Fakhari, Khodadadi, & Busemeyer, 2018). This raises the question of what conditions lead to the use of each type of learning process. On the one hand, the simple learning of Q-values for each state and action by model-free processes allow it to avoid the complexity of solving the backward equations required by the model-based method. On the other hand, the model-free process is inflexible for adapting to changing conditions, whereas the model-based process can adapt quickly based on a representation of the problem structure. In short, model-based processes provide greater accuracy with more effort, and model-free are less accurate but also require less effort.

9.4 Strategy Learning

It has been argued that rather than learning a specific action to take at each state of a decision, people learn entire strategies rather than simple actions (Erev & Barron, 2005). Several models for learning strategies have been developed with different types of applications in mind.

Busemeyer and Myung (1992) developed a model for learning strategies to categorize objects in signal detection tasks. A strategy in this application consists of a rule (i.e., a set of procedures) for categorizing a stimulus. For example, when presented with a two-dimensional stimulus, one type of rule might be to use a single linear boundary; another type of rule might be to use an elliptical boundary. An important feature about rule learning is that each qualitatively different rule also involves a set of continuously varying parameters that affect the performance of the rule. For example, a linear rule involves learning a slope and intercept; an elliptical rule involves learning a different set of parameters. Therefore, this model involves two types of learning mechanisms: one is a reinforcement learning process for learning qualitatively different rules (e.g., linear versus elliptical), and a second is a hill-climbing process that learns the parameters of a rule. These two learning processes work in tandem to generate the integrated learning process. They showed that their proposed reinforcement learning model outperformed a fixed, small-sample model for learning qualitatively different rules, and the hill-climbing algorithm outperformed earlier criterion-learning models for signal detection (Dorfman & Biderman, 1971).

Rieskamp and Otto (2006) developed a model for learning strategies for making probabilistic predictions based on searching and evaluating a set of cues with varying cue validities. They proposed that people learn qualitatively different heuristic rules, such as "take the best," which is a lexicographic rule, as opposed to a weighted additive rule (weighted sum of cue values, with weights equal to cue validities) for making predictions. The basic idea is that each application of a strategy produces an experienced payoff, and a reinforcement learning algorithm is applied to learn the expected payoff of a strategy. Finally, the expected payoff based on this learning is used to select a strategy using a soft max/ratio of strengths type of rule. They showed that their strategy learning model outperformed an exemplar learning model (Juslin, Olsson, & Olsson, 2003) for these types of tasks.

Most recently, Lieder and Griffiths (2017) proposed what they call a resource rational model of strategy learning and selection, based on a theory of rational metareasoning. According to their model, people learn to efficiently choose the strategy with the best cost-benefit trade-off by learning a predictive model of each strategy's performance. Like Payne, Bettman, and Johnson (1993), they argue that the cost associated with a strategy includes not only the losses produced by the payoff of using a strategy, but also the computational cost (approximated by execution time) of executing the strategy. Following earlier work on learning with information-processing tasks, a key feature of this theory is to carefully model the computational costs of executing a strategy in a given environment. A new feature of this theory is that it proposes that a person learns how to map a feature vector that describes the current problem into a prediction of cost-benefit value of each strategy, which then can be used to select a strategy that maximizes the expected cost-benefit trade-off. This capability to learn to predict the cost-benefit of a strategy for a given problem provides this theory with the ability to account for aspects of strategy leaning that were left unexplained by previous theories.

9.5 Combining Reinforcement Learning and Sequential Sampling Models

On the one hand, reinforcement learning models have relied on simplified choice processes, such as the soft max/ratio of strengths rule. They do not provide any mechanisms to account for the within-trial dynamics that lead to a single choice. Consequently, they are unable to account for choice time or the effects of time pressure on choice. On the other hand, the sequential-sampling models described in chapter 8 provide a dynamic within-trial mechanism that accounts for both choice probability and response time. However, they fail to describe the across-trial dynamics produced by learning from trial-by-trial feedback. To combine these two approaches to decision-making and learning, several new attempts have been made to integrate a diffusion-model choice process with reinforcement learning models, thereby capturing both within- and across-trial dynamics. The basic idea is to use the *Q*-values that represent the estimated preferences for each action to determine the mean drift rate of a diffusion model (Pedersen, Frank, & Biele, 2017; Fontanesi et al., 2019). Turner (2019) proposed a model that formally connects the ideas of instance-based learning and model-free types of reinforcement learning, all within the same sequential sampling model, and also explicitly tests the utility of one type of representation over another based on trial-by-trial decisions (i.e., choice and response time). So far, these models have been applied only to repeated-choice, two-armed bandit problems. Decision-making with MDPs involve planning, which might require more complex within-trial planning processes that would affect decision time.

9.6 Conclusions

Research on experience-based decisions has been growing rapidly over the past couple of decades. This was partly stimulated by the description-experience gap, which demonstrated preference reversals between choices based on description rather than those based on experience. Another reason for the rising interest in experience-based tasks is the connection between the learning algorithms proposed for these tasks and neuroscience, which is described in the next chapter. This new line of research on experience-based decision tasks includes not only simple, multiarm bandit–type problems, which are finished after a single decision, but also more complex Markov decision problems, which require planning across a sequence of decisions.

For both types of problems, learning models are combined with choice models to predict the trial-by-trial choices produced with these tasks. It is interesting to note that although fairly sophisticated learning processes have been proposed, the choice processes are usually based on the very simple Luce choice model, which has been criticized in previous chapters. What needs to be done in future research is to integrate more sophisticated, sequential sampling choice models with experience-based learning models so that the resulting models are capable of predicting not only choice, but also decision time.

9.A Appendix

Here is an example program used to predict choices in the IGT, which is available at https://github.com/jbusemey/ChoiceModeling.git. The first part reads the data and sets the parameters. The loop computes the predictions for the learning model.

```
function chi=Fitr1(parm,data)

chi=0 ; %initialize chi square
v=[0;0;0;0]; %initialize expectations for each deck

nt=200; % no of trials
y=data(:,1); % choice response
x=data(:,2:4); %payoffs received

a=exp(parm(1)); % learning rate
a=a./(1+a);

b1=exp(parm(2)); %attention go losses versus gains
b1=b1./(1+b1);

c=exp(parm(3)); % exploration/exploitation sensitivity

%-----------------------------------MODEL
```

```
%begin of training
for tt=2:nt
  t=tt-1;
  yt=y(t,:) == [1 2 3 4]'; % current choice just made this
  choice
  ytt= y(tt,:)== [1 2 3 4]'; % the next choice one step ahead
  we wish to predict
  xt=x(t,:); % current payoff just experienced; one row in x,
  for trial t

  win=abs(xt(2))/100; % rescale payoffs
  loss=abs(xt(3))/100; % rescale payoffs

  v=v+a * yt.* ((1-b1)*win - b1*loss - v); % revise expectations
  gam=(t/10)^c; % sensitivity

  s=exp(gam*v)+ 0.0000000001; % strength
  p=s./sum(s); % prob of choice for each deck

  pp =.0001 +.9998*p; %bounded between .0001 and .9999
      % to avoid log of zero
  chi=chi+log(pp)'*ytt;

end
%end of training
%----------------------------------------MODEL---------------
-------------------
chi=-2*chi; % -2 times log likelihood
```

10 Decision Neuroscience of Choice

The primary topic of this book has been behavioral models of choice, but it is important to consider how findings from a field called *decision neuroscience* relate to these choice models. There is a massive body of literature in this field and the closely related field known as *neuroeconomics* (see Glimcher & Fehr, 2014). We don't intend to give a comprehensive review of this large body of work. Instead, we limit our review to neuroscience findings that are directly related to several of the previously discussed choice models, including expected utility models (chapter 2), signal detection models of categorization (chapter 5), accumulation to threshold choice models (chapter 6), and reinforcement learning models (chapter 9).

10.1 Neural Substrates of Expected Utility Theory

Researchers in the field of neuroeconomics have investigated the neural substrates underlying the basic elements of expected utility theory in humans using functional magnetic resonance imaging (fMRI) technology. Essentially, participants are asked to perform decision tasks, such as risky decisions or delayed discounting decisions, while inside an MRI scanner. Brain activation is indirectly monitored by blood oxygenation level–dependent (BOLD) signals produced during the task. This technology provides a relatively precise identification of the location or region of brain activation. However, the technology does not provide good resolution of the timing of events because the observed BOLD signal is a very slow, temporally filtered signal of the underlying brain activity. Sometimes electroencephalography (EEG) is used in conjunction with fMRI, which has better time resolution but poorer spatial resolution.

One of the basic assumptions of expected utility theory is that risky options, delayed consequences, and consumer options are all mapped into a single variable called *utility*. There is now a substantial amount of evidence collected via fMRI that there is a utilitylike signal that appears in two locations in the brain: the ventral striatum, located in a primitive part of the brain called the *basal ganglia*; and the ventral medial prefrontal cortex (VMPC), which is a more recently evolved and advanced part of the brain. Both regions combine the outputs of many antecedent brain areas onto a single common scale that is appropriate for direct comparison and choice. They are also connected with many areas involved in affect and cognition, such as the hippocampus, amygdala, hypothalamus, and lateral prefrontal cortex. In this section, we focus on the VMPC, and the next section discusses the striatum.

A good example of a study on the neural basis of utility is a study by Kable and Glimcher (2007), who investigated preferences for an immediate small gain over a larger amount received after a longer delay, and found that BOLD activity in the VMPC in each participant was strongly correlated with the discounted value behaviorally estimated for each delayed reward. Similar results were found with risky choices (Tom et al., 2007). Meta-analysis studies of this vast body of literature consistently demonstrate that activity in these areas correlates with human preferences at a within-subject level for a wide range of choices, including food and drink rewards, monetary rewards, gains, losses, social rewards, and even more abstract rewards (Levy & Glimcher, 2012). The fact that the same areas are involved in appetitive and aversive consequences runs contrary to the idea that there are separate appetitive and aversive valuation systems in goal-directed choice, but this is still debatable.

As pointed out by Rangel and Clithero (2014), additional evidence is needed to conclude that activity in the VMPC represents a value signal that guides preference. Supporting the latter inference, additional research has found that this activation (1) does not simply reflect factors such as arousal or motor preparation, (2) is not confounded with prediction errors, (3) is not produced by action-related information, (4) anticipates rather than depends on the final outcome of the decision process, and (5) serves as inputs to comparison processes in the areas of the dorsal medial prefrontal cortex and the intraparietal sulcus, which are involved in implementing a comparison process that results in choice. These five conclusions provide strong evidence supporting the hypothesis that VMPC activation represents a signal used to compute preferences. Perhaps the strongest evidence for linking neural activity in the VMPC to choice behavior was obtained by manipulating the activity in this region at the time of choice using transcranial magnetic stimulation—this neural change produced a corresponding change in choice behavior (Baumgartner et al., 2011). What remains relatively unknown are details about how these signals drive action neurons in the frontal and parietal cortices that finally produce a choice.

10.2 Multiple Neural Substrates of Category Learning

Multidimensional signal detection models have been used to describe how people categorize stimuli, such as how a radiologist categorizes a medical image as indicating the presence or absence of a disease based on features such as size and density of a node. Many studies have examined how people learn categorization rules in these tasks, and this research suggests that it is important to distinguish between two types of tasks: rule-based and information integration tasks (Ashby & Maddox, 2005). Figure 10.1 illustrates two kinds of two-dimensional signal detection models (also known as *general recognition models*): an integration task and a rule-based task.

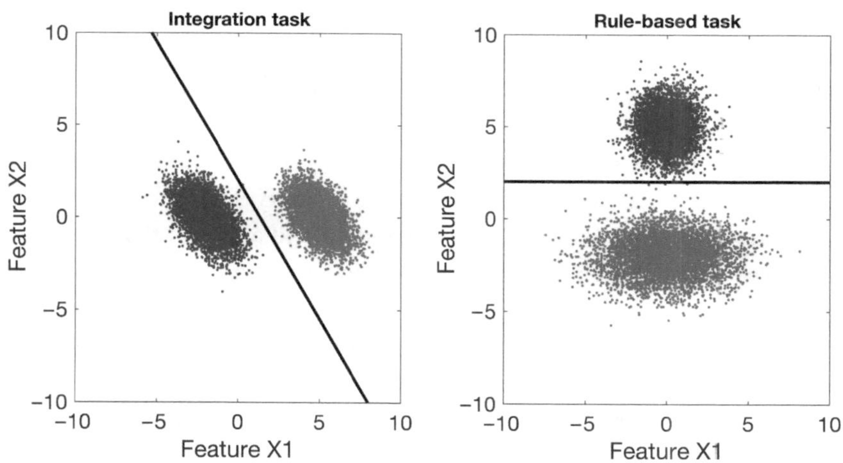

Figure 10.1
Examples of integration and rule-based tasks.

The categories can be learned via a logical reasoning process for rule-based tasks, such as tasks in which only one stimulus dimension is relevant and the participant's task is to discover the relevant dimension. Rule-based tasks are easy to describe verbally. A wide body of evidence suggests that rule-based tasks are learned by a declarative memory system that relies on working memory and executive attention in a more advanced part of the brain, such as the dorsolateral prefrontal cortex. For information-integration tasks, the categories must be learned by integrating information from two or more incommensurable stimulus components using some form of weighted linear combination of dimension values. Information-integration tasks are difficult to describe verbally. Evidence suggests that such tasks rely on procedural learning that is mediated largely within a more primitive part of the brain in the basal ganglia called the *striatum*.

The declarative rule-based system is restricted to learning only a small set of category structures that satisfy a logical rule, but it does so quickly. The procedural system can learn a much larger set of category structures, but it does so in a slow, incremental fashion and is highly dependent on reliable and immediate feedback. The procedural and declarative memory systems compete for control of motor resources. In fact, neuroimaging studies have found an antagonistic relation between the two systems: they learn simultaneously, but if the rule-based system succeeds, then the declarative system inhibits the procedural system. This two-system category learning process is described in detail by a computational neural model called COVIS (Ashby et al. 1998), which stands for "Competition between Verbal and Implicit Systems."

The rule-based system operates with two separate subnetworks—one that maintains candidate rules in working memory, tests those rules, and mediates the switch from one rule to another; and a second that generates or selects new candidate hypotheses. Long-term representations of each possible salient rule are encoded in neural networks in the sensory association cortex. These cortical units send excitatory signals to working memory units in the lateral prefrontal cortex, which send recurrent excitatory signals back to the same cortical units, thereby forming a reverberating loop. At the same time, the prefrontal cortex is part of a second excitatory reverberating loop through the medial dorsal nucleus of the thalamus. These double reverberating loops maintain activation in working memory units of the prefrontal cortex during the hypothesis-testing procedure. The process of switching attention from an incorrect rule is initiated by a reduction in the excitatory input in the prefrontal cortex to the head of the caudate nucleus, which in turn increases inhibition of the thalamus and breaks the cortical-thalamic loop (Alexander, DeLong, & Strick, 1986). Then the anterior cingulate selects among alternative rules by enhancing the activity of the specific working memory unit in the prefrontal cortex that represents a particular rule. The correct rule is eventually transferred to long-term memory by way of projecting from the prefrontal cortex to the hippocampus.

The procedural-learning system learns arbitrary stimulus-response associations in a slow, incremental manner using a dopamine-mediated reinforcement learning process. Procedural learning in the striatum is facilitated by a dopamine-mediated reward signal from the substantia nigra pars compacta. Positive outcomes increase phasic dopamine levels, which in turn strengthens recently active synapses, whereas negative outcomes decrease dopamine levels to below baseline, which weakens recently active synapses. The stimulus-response associations are formed within the striatum, which includes the caudate nucleus and the putamen. The striatum sends projections to several prefrontal and premotor cortical areas. The posterior regions of the putamen project primarily into the motor and premotor areas of cortex. In contrast, the anterior putamen and caudate project to the prefrontal regions of the cortex.

The COVIS model makes several strong predictions about learning in rule-based versus information-integration tasks, which have been supported by experiments. For example, the model predicts that because the declarative system has access to working memory and executive attention, it should be relatively unaffected by changes in the timing and form of the feedback signal. In contrast, the procedural system is highly sensitive to these feedback parameters because a dopamine-mediated reward signal is necessary for learning to occur in the striatum, the absence or delay of such a reward signal should greatly interfere with this form of category learning. In support of these predictions, (1) people can learn rule-based but not information integration categories in the absence of any trial-by-trial feedback; (2) information integration, but not rule-based learning, is worse with observational training than with feedback training; (3) information integration, but not rule-based learning, is impaired with feedback delays; and (4) information integration, but not rule-based learning, is impaired if the feedback is batch delivered (Maddox & Ing, 2005).

10.3 Neural Mechanisms of Evidence-Accumulation Processes

Sequential sampling models assume that evidence for each action is sequentially sampled and accumulated over time until the accumulation for one action reaches a criterion level of strength to make a choice. Although these models were initially formulated as behavioral models of choice and decision time for perceptual and memory tasks (see chapter 6), it was later discovered that they also have fairly direct connections with neuroscience findings concerning the accumulation of neural activation that gives rise to motor action (for reviews, see Shadlen & Kiani, 2013; Gold & Heekeren, 2014).

Much of the research connecting sequential sampling models to decision neuroscience was initiated by Newsome and colleagues (1989) using a dot-motion paradigm. This paradigm is useful for identifying the cortical mechanisms of perceptual decisions about the direction of visual motion. In a dot-motion task, the stimulus is a display of jiggling dots, with a proportion jiggling in a specific direction (say, the right) and the remainder jiggling in random directions. The proportion jiggling to the right (in this example) is called the *coherence* of the stimulus. The task is to decide the direction of coherent motion and indicate its direction by making a saccadic eye movement toward a visual target in that direction. If the eye movement is in the correct direction, then a reward is delivered. Manipulating the percentage of coherently moving dots provides a precise way to control the average strength of evidence provided by the dynamic stimulus.

Newsome and colleagues (1989) used electrophysiological recording of visual neurons in the medial temporal area of monkeys to relate neural activation to behavioral decision-making. The medial temporal area (also known as *area V5*) is tuned toward motion detection. Newsome and colleagues found that the strength of spiking activity favoring one direction over another in the medial temporal area was directly increased by the coherence of the dot motion. Furthermore, electrical stimulation of the medial temporal area favoring one direction directly increased the probability of a response in that direction, revealing a causal connection between activation and behavior. The medial temporal area provides a fairly static distribution of activity that depends on coherence, but it provides no accumulation of evidence across time. The activity in the medial temporal area reflects the moment-by-moment motion evidence, and the activation by a stimulus tends to stay at a level that reflects the coherence of the stimulus.

To investigate the accumulation of evidence, electrophysiological activity needed to be recorded from premotor neurons in monkeys, such as in the lateral intraparietal area (Shadlen & Newsome, 1996, Gold & Shadlen, 2007) or the frontal eye fields (Hanes & Schall, 1996), which receive inputs from the visual neurons and then later drive the saccadic eye movement system. It turns out that the activity recorded in these premotor neurons accumulates over time at a rate that is proportional to the strength of activation in the medial temporal area. Increasing activation in the premotor area also produces increasing choice probability in the monkeys. Furthermore, the time of choice was determined by the time required for activation in the premotor neurons to reach a hidden threshold, as illustrated in figure 10.2 (Roitman & Shadlen, 2002).

These findings connecting neuroscience to choice behavior support several important principles (Shadlen & Kiani, 2013). The neural decision system (1) is flexible with respect to timing of decisions, (2) involves the integration and accumulation of evidence from multiple modalities, (3)

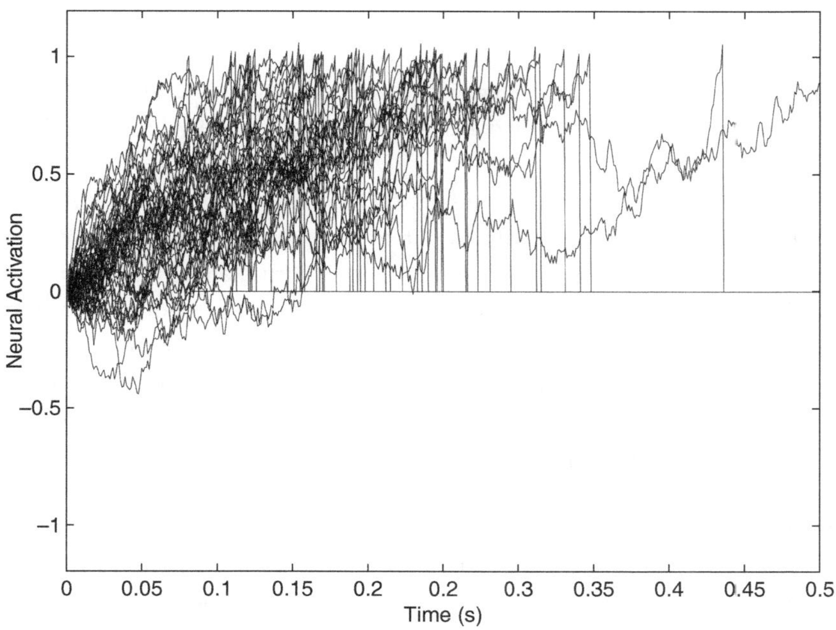

Figure 10.2
Simulation of neural activation accumulating up to a hidden threshold, which then produces a saccadic eye movement in one direction.

provides a probabilistic representation of the accumulated information, (4) computes a neural signal that is a direct representation of a decision variable, (5) produces a continuous flow from sensory evidence to behavior activation, (6) provides a stopping rule for terminating evidence accumulation, and (7) serves provisional intentions rather than building an abstract conceptual representation.

The evidence accumulation process has also been investigated in the human brain using fMRI. Accumulation of evidence was reflected in BOLD activity in the dorsolateral prefrontal cortex (Heekeren et al. 2004; Philiastides & Heekeren, 2009). Using the dot-motion task with humans, Liu and Pleskac (2011) found neural signatures of evidence accumulation—a higher fMRI response on low coherence trials than high coherence trials—primarily in saccade-related sensorimotor areas, such as the frontal eye fields, and nonsensorimotor areas of the anterior insula and inferior frontal sulcus. Mulder and colleagues (2012) found evidence that the orbitofrontal cortex is involved in generating decision bias, based on the correlation between orbital frontal cortex activation and the accumulation model bias parameter in various tasks. Using a preferential choice task, Mohr, Heekeren, and Reiskamp (2017) found that accumulated evidence in favor of the chosen option was positively correlated with brain activity in the medial orbitofrontal cortex and negatively correlated with brain activity in the anterior insula.

EEG has also been used to investigate evidence accumulation in humans. Studies of the lateralized readiness potential (a voltage difference between event-related potentials over the motor cortex contralateral and ipsilateral to the effector) revealed partial activation in the movement preparation stage (Osman et al., 1992) indicating a continuous flow of activation from sensory to motor systems. The late temporal component of event-related voltage potential EEG was found to be correlated with mean drift rate in a diffusion model (Ratcliff, Philiastides, & Sajda, 2009), suggesting that the late temporal voltage component reflects the postsensory evidence that is fed into the accumulation process.

One of the primary goals of relating neuroscience to decision-making is to use the constraints imposed by neural data to determine the mechanisms underlying decision-making (Ratcliff et al., 2007). Purcell et al. (2010) used the firing rate of a specific neuronal population in monkeys as the input to an accumulator network. Using their model analyses, they came to three key conclusions: (1) accumulator models that used visual neuron activity as input were sufficient to account for observed variability in behavior during saccade related visual search tasks; (2) although many models make indistinguishable predictions of behavior, they make different predictions about the characteristics of neural activity; and (3) to fully account for both behavioral and neural data, the accumulator models must assume that the flow of perceptual evidence to the accumulator is gated such that evidence from the visual neurons below a cutoff is not passed to the accumulator in the motor neurons. More recently, Turner, van Maanen, and Forstmann (2015) used prestimulus human brain activity measured by fMRI to constrain the parameters of an accumulation model of choice and response time, and they showed that this enhanced model could predict accuracy and response time from single trials better than models using behavior alone.

10.4 Dopamine and Reinforcement Learning

Many reinforcement learning models use an error signal to update the expectation for an action based on the discrepancy between the predicted and obtained outcomes of an action (Dayan & Abbot, 2001). Neuroscience research has established a strong link between the error signal from reinforcement learning models and a neurotransmitter called *dopamine* found in the midbrain (Daw & Tobler, 2014; Niv, 2009). It should be noted, however, that dopamine largely has two functions: phasic dopamine signals serve reinforcement learning, whereas tonic dopamine levels in the striatum facilitate movement (which connects it to Parkinson disease). Here, we focus on the phasic function of dopamine.

The majority of dopamine neurons are found in three contiguous nuclei called the *retrorubral nucleus,* the *substantia nigra pars compacta,* and the *ventral tegmental area.* The dopamine neurons from these nuclei send projections to many regions, such as the striatum, the amygdala, and the frontal cortex. These projections are used to broadcast a prediction error signal to the synapses involved in learning stimulus–response relations. The dopamine neurons are coupled to one another to provide a high degree of interneuronal synchrony, which allows the target synapses to receive a single coherent error signal. Dopamine alters the strength of synapses: high levels facilitate increases in synaptic strength, and low levels have the opposite effect. Early researchers associated dopamine with motivation—they thought that it served as the brain's pleasure reward system. Many drugs that are abused activate and/or maintain the activation of dopamine neurons. A more modern view is that, rather than producing subjective pleasure, dopamine provides a reinforcing function to increase the response probability for rewarded actions. Note, however, that the function of dopamine is not a simple signal of reward; it depends on prediction error rather than reward magnitude (Schultz, Dayan, & Montague, 1997).

One of the first investigations of the response of dopamine neurons to prediction error was performed using electrophysiological recordings in monkeys. Figure 10.3 illustrates the basic idea. The monkeys were trained to predict a reward that occurred two seconds following a randomly presented cue. Early in training, the reception of an unexpected reward elicits a prediction error and concomitant dopamine response. After learning occurs to associate the reward with the cue, the dopamine response comes

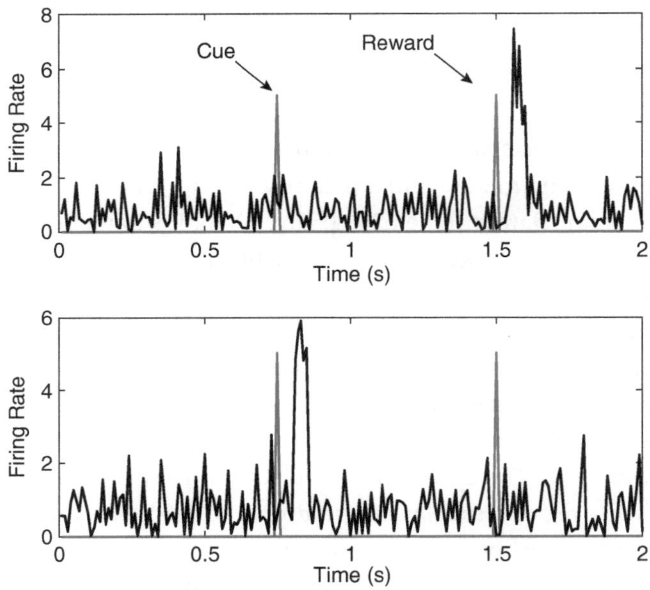

Figure 10.3
Simulation of dopamine activation to a cue and a reward. The first pulse in each figure represents the presentation of a cue, and the second pulse represents the presentation of a reward. The jagged curve in each figure represents the firing rate. The top figure shows the results for early training, and the bottom figure shows the results for later training.

after the unpredictable cue before the reward. No response occurred after the reward if it occurred as predicted by the cue. However, if no reward was received after a cue that predicted it, then the dopamine response was suppressed. Shortly afterward, theoreticians recognized the close relation between these results and the error signals used in reinforcement learning models (Montague et al., 1994, 1995; Schultz, 1998)

As an additional demonstration of the importance of prediction error, a later study manipulated the probability of reward (Fiorillo, Tobler, and Schultz, 2003). Dopamine responses were stronger for rewards when there was a low probability of reward. Moreover, when rewards failed to occur, dopamine neurons showed a decrease in firing at the time that the reward would have been expected. Moreover, detailed quantitative predictions derived from a temporal-difference learning model were supported by the dynamic response of dopamine to a training cue (e.g., Bayer & Glimcher, 2005). Later studies also supported these findings with humans using model-based fMRI technology (O'Doherty et al., 2003).

Jessup, Busemeyer, and Brown (2010) obtained results indicating that the anterior cingulate cortex (ACC) produces a signal for unexpected outcomes, much like a prediction error signal. However, the ACC—a proximal and somewhat overlapping area of the VMPFC—is also thought to serve higher-order cognitive functions, such as executive control and planning (in other words, the sorts of things that may be involved in goal-directed learning).

Analogous to the multiple systems used in category learning, reinforcement learning research suggests that the brain uses two learning systems to perform sequential or multistage decision tasks (Daw & O'Doherty, 2014): a habitual system, which learns to repeat previously successful actions; and a goal-directed system, which evaluates actions on the basis of their specific anticipated consequences. The habitual system corresponds to model-free reinforcement learning (described in chapter 9), which is incapable of evaluating novel actions or reevaluating previous experiences. The goal-directed system corresponds to model-based reinforcement learning (also discussed in chapter 9), which forms a representation of the environment and evaluates actions more prospectively.

A key method for distinguishing the systems is to examine abrupt changes to reward contingencies: the habitual system slowly adapts to such changes, whereas the goal-directed system uses the representation of the environment to adjust actions quickly. The habitual system involves sensorimotor and cortical inputs to the more primitive dorsolateral striatum. The habitual system (model-free learning system) is also the system that is updated using error-driven learning associated with dopamine activation. The goal-directed system in humans involves links between the more recently evolved and advanced medial prefrontal cortex and dorsal medial striatum. The hippocampus may also be involved in goal-directed learning by providing an internal model of the environment. Both systems are innervated by dopaminergic inputs from the ventral tegmental area and substantia nigra pars compacta.

For example, Hampton and colleagues (2006) was performed a study that implicated the VMPC in the goal-directed (i.e., model-based) learning system. Participants engaged in probabilistic reversal learning, in which selection of one action led to a high probability of monetary gain, while the other action leads to a high probability of monetary loss. After a period of time, the contingencies reversed. A model-based learning algorithm that incorporated knowledge of the reversal structure of a task predicted performance better than a model-free algorithm that did not. Moreover, using fMRI technology, the VMPC activation was found to correlate better with value signals generated by the model-based algorithm than by the model-free algorithm.

The distinction between habitual and goal-directed systems may also have some connection to judgment and decision-making ideas that stress the distinction between controlled or deliberative mode, and an automatic or intuitive mode for making decisions (Kahneman, 2003; Shiffrin and Schneider, 1977). As with the multiple systems of category learning, the habitual and goal-directed systems compete to control action selection. Problems sometimes arise in situations where habits come to dominate behavior, which has implications for neuropsychological models of addiction and other psychiatric disorders involving compulsive behavior (Everitt & Robbins, 2005).

10.5 Final Comments

There are several convincing reasons for trying to integrate cognitive choice models with decision neuroscience. Ashby and Helie (2011) list the following reasons. First, understanding decision neuroscience underpinnings can increase the number of constraints on behavioral models. For example, if neuroscience data implicate the hippocampus in a behavior, then models of this behavior must all share some hippocampal-like properties.

Second, attending to the neuroscience data can expose relationships between seemingly unrelated behaviors. For example, neuroscience might identify similar neural circuits underlying seemingly unrelated tasks, which raises the possibility that they share some previously unknown but deep functional similarity.

Third, in many cases, studying the underlying neuroscience leads to surprising behavioral predictions that would be difficult or impossible to derive from a purely cognitive approach. Finally, integrating cognition with neuroscience allows a converging operations approach to model testing using both behavioral and neuroscience data, including single-unit recording data, fMRI data, and EEG data.

In the past, computational neuroscientists have tended to use fits to behavioral models to then predict neural data (e.g., model-based fMRI analyses). But recently, work is being done (e.g., Turner et al., 2017, 2019) that integrates both the behavioral and the neural, so that both measures can serve to constrain the parameters and test the models.

11 Quantum Choice Models

Why is choice probabilistic? One idea is that choice is fundamentally deterministic—each choice is determined by a real preference state that exists at the moment immediately before a choice is made, but the preference state changes across occasions. Another idea is that choice is fundamentally indeterministic (Glimcher, 2005)—there is no real preference state existing at the moment immediately before a choice is made; instead, the choice is essentially probabilistic.

A random utility model (discussed in chapter 3) is actually deterministic because choice is based on a specific state that exists when the choice is made. Choice probability arises from the theorist lacking full information about this precise state.

Recently, interest has been growing in the use of principles based on quantum probability theory, which posits that choice is essentially indeterministic in nature (Mogiliansky, Zamir, & Zwirn, 2009). The reason for considering a quantum approach is that choice probabilities sometimes violate properties that lie at the heart of classic random utility models (Aerts, 2009; Khrennikov, 2010; Busemeyer & Bruza, 2012; Yukalov & Sornette, 2011).

Quantum cognition is a rapidly growing field (Pothos & Busemeyer, 2013), and we do not attempt to review all the research in this area here. Instead, we provide a brief introduction to the basic principles, some empirical motivation for considering a quantum approach, followed by a detailed application to a puzzling finding from the field of decision-making regarding a violation of one of the fundamental axioms of rational decision-making called *dynamic consistency* (Busemeyer, Wang, & Shiffrin, 2015).

11.1 Basic Principles

Why apply quantum theory to choice behavior? Classical probability theory evolved over several centuries, but an axiomatic foundation for it did not exist until Kolmogorov (1933) provided one. Quantum mechanics was invented by a brilliant group of physicists in the 1920s, but its status as a general theory of probability became clear only once the axiomatic foundation was provided by Dirac (1930) and von Neumann (1932). Quantum probability theory may be more suitable than classical theory to describe human behavior in many situations (Bruza, Wang, & Busemeyer, 2015). For example, human judgments are not commutative and order effects are pervasive; human decisions often violate predictions based on the law of total probability. Violations of classical axioms are not the only reasons for considering a quantum framework. Principles from quantum theory resonate with deeply rooted psychological conceptions: *superposition* captures the intuitive feelings of conflict, ambiguity, and uncertainty of mental states, while *complementarity* captures the idea that measurement constructs rather than records a mental property.

What is quantum probability theory? To keep this presentation simple, we assume finite spaces, although both classical and quantum probability theories can be extended to infinite spaces (for more detail, see Busemeyer & Bruza, 2012). Classical theory begins by postulating a set called the *sample space*, Ω, a set of elements that contains all the events, and this set has cardinality N. Quantum theory begins by postulating a vector space (technically called a *Hilbert space*), V, which contains all the events, and this vector space has dimension N. Classical theory is based on the premise that an event, such as A, is a subset $A \subseteq \Omega$ of the sample space. Quantum theory is based on the premise that an event, such as A, is a subspace $A \subseteq V$ of the vector space. Each subspace, such as for example the event A, corresponds to a projector P_A that project vectors onto the subspace. Classical theory postulates a state represented by a function $p: \Omega \rightarrow [0,1]$, which assigns probabilities to mutually exclusive events in an additive manner. Quantum theory postulates a state represented by a unit length vector $\psi \in V$, which assigns probabilities to orthogonal events in an additive manner using the following algorithm: $q(A) = ||P_A \psi||^2$, where P_A projects vectors onto the subspace A. Classical theory defines a conditional probability of event B given A as $p(B|A) = p(A \cap B)/p(A)$, so that the joint probability is defined by $p(A \cap B) = p(A)p(B|A) = p(B)p(A|B) = p(B \cap A)$, and order does not matter. Quantum theory defines this conditional probability as follows:

$q(B|A) = ||P_B P_A \psi||^2 / q(A)$, so the sequential probability is defined by $q(A, B) = q(A)$ $q(B|A)$, which is not necessarily the same as $q(B)q(A|B) = q(B, A)$, and the order matters.

Classical probability theory can be dynamically extended to evolution across time by a Markov process that evolves state p according to the transition operator T, which preserves the total probability of the state $p(t) = T(t) \cdot p(0)$. Quantum theory can be extended to evolution across time by a quantum process that evolves state ψ according to a unitary operator U that preserves the length of the state $\psi(t) = U(t) \cdot \psi(0)$.

The quantum principles are illustrated by the following demonstration of a violation of total probability in a three-dimensional space. Suppose that a person is asked to judge her degree of belief about a defendant's guilt on a three-level scale (Low, Medium, or High) and/or judge whether the punishment for the crime should be severe or not. Figure 11.1 illustrates the computation of the probability of assigning severe punishment under two conditions: (a) the person judges whether the defendant has high guilt or not (not high means the judgment is low or medium), and then decides whether severe punishment should be assigned to the crime; (b) the person only decides whether the crime should be assigned severe punishment (without being asked to judge guiltiness).

The ray spanned by the vector H represents the event high guilt, and the plane spanned by the pair of vectors L and M represents the event not high guilt (low or medium guilt). The ray spanned by the vector S represents the event severe punishment (not severe is the orthogonal complement to S). The vector labeled *Psy* represents the person's state for making judgments.

The projection, R, of the state *Psy* onto the ray spanned by H determines the probability of judging high guilt: the probability of high guilt is the squared length of R, which equals .091. The projection, T, of the state *Psy* onto the plane spanned by the vectors L and M determines the probability of not high (low or medium) guilt: the probability of not high guilt (low or medium) equals the squared length of T, which equals .909.

The probability of judging high guilt and then assigning severe punishment is obtained by first projecting *Psy* onto H to reach R, projecting the result R onto S to produce projection U, and then computing the squared length of U, which equals .045. (Follow the arrows starting from Psy and then pointing left and then down.)

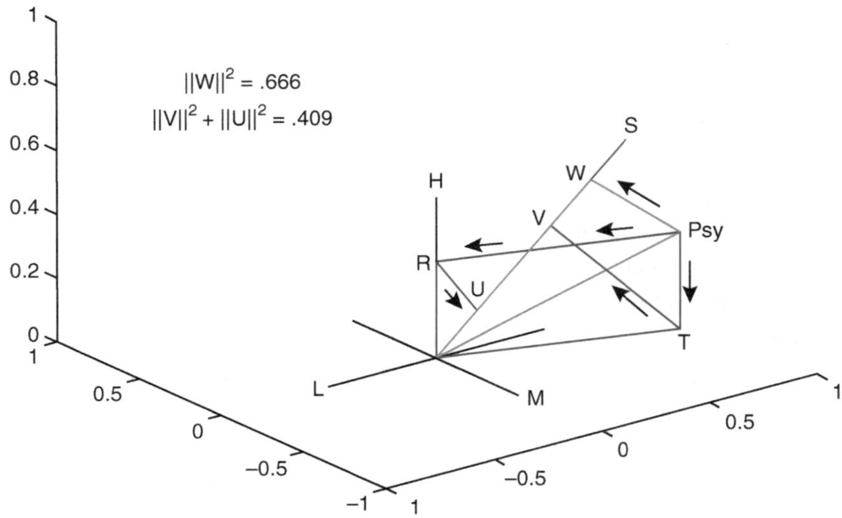

Figure 11.1
Demonstration of a violation of total probability. The total probability of Severe punishment (after first measuring belief about guilt) is computed by summing the squared lengths of projections *U* and *V* on the ray spanned by *S*. The probability of Severe punishment (without measuring belief of guilt) is computed by the squared length of the projection *W* on the ray *S*.

The probability of judging not high guilt and then assigning severe punishment is obtained by first projecting *Psy* onto the *L*, *M* plane to reach *T*, projecting the result *T* onto *S* to produce projection *V*, and then computing the squared length of *V*, which equals .364. (Follow the arrows starting from *Psy* and then pointing down and then up.)

The probability of severe punishment without measurement of guiltiness is obtained by projecting state *Psy* directly onto vector *S* to produce projection *W* and the squaring its length, which equals .666. (Follow the arrow starting from *Psy* and then pointing up.)

Thus, we see that this simple model produces a violation of total probability. The total probability of severe punishment, following measurement of guiltiness equals the length of *U* squared (.045) plus the length of *V* squared (.364), which equals .409. However, this is less than the probability of assigning severe punishment alone .666 (when not asked about guiltiness).

This same example also produces order effects. The probability of choosing severe punishment and then high guilt equals .333, but the probability of choosing high guilt and then severe punishment equals only .045. The order effect occurs because projecting first on severe punishment produces a large projection but projecting first on high guilt produces a very low projection.

11.2 Violations of Predictions from the Law of Total Probability

At this point, the reader might question why we need a theory that can account for violations of predictions from the law of total probability. When has this happened? Actually, there are many examples (see Bruza, Wang, & Busemeyer, 2015, for discussion), but one example is provided by a well-known experimental finding called the disjunction effect. Tversky and Shafir (1992) designed a study to test an axiom of utility theory called the sure thing principle (Savage, 1954). According to this principle, if you prefer action A over B under state of the world X, and you also prefer A over B under the complementary state not-X, then you should prefer A over B when the state is unknown. They tested this principle by using a simple two-stage gambling experiment: participants were told that they had just played a gamble (even chance to win \$200 or lose \$100), and then they were asked to choose whether to play the same gamble a second time. In one condition, they were informed that they won the first play; in a second condition, they were informed that they lost the first play; and in a third condition, they did not know the outcome. Surprisingly, the results produced what they called a disjunction effect: following a win, they chose to take again on 69 percent of the trials; following a loss, they chose to take again on 59 percent of the trials; but when the outcome was unknown, they chose to take again for only 36 percent of the trials.

This is a violation of a prediction from the law of total probability for the following reason. The probability of taking the gamble in the unknown condition is expected to be equal to

$$p(take \mid unknown) = p(win) \cdot p(take \mid win) + p(lose) \cdot p(take \mid lose).$$

Therefore, the probability to take the gamble for the unknown condition should lie in between $p(\text{Take} \mid \text{win})$ and $p(\text{Take} \mid \text{loss})$, but in fact, $p(\text{Take} \mid \text{unknown})$ falls below both of these, violating this prediction. Tversky and Shafir (1992) also concluded that their results showed evidence for violations of the sure thing principle, because many participants chose to play again when the first stage was known to be a win, and also when the first stage was known to be a loss, and then switched and decided not to play when the outcome was unknown.

Pothos and Busemeyer (2009) provided a simple account for these results using a quantum choice model. The basic idea is the following. Define ψ as the state under the unknown condition. Define P_T as a projector that represents the decision to take the gamble on the second stage. Define P_W as the projector for the event winning the first stage and $P_L = I - P_W$ as the projector for the event losing the first stage (I is the identity matrix). Then the probability to take the gamble on the second stage can be expressed as

$$\begin{aligned} q(take \mid unknown) &= \|P_T \cdot \psi\|^2 = \|P_T \cdot I \cdot \psi\|^2 \\ &= \|P_T \cdot (P_W + P_L) \cdot \psi\|^2 = \|P_T \cdot P_W \cdot \psi + P_T \cdot P_L \cdot \psi\|^2 \\ &= \|P_T \cdot P_W \cdot \psi\|^2 + \|P_T \cdot P_L \cdot \psi\|^2 + Int \\ &= q(win) \cdot q(take \mid win) + q(lose) \cdot q(take \mid lose) + Int \end{aligned}$$

where *Int* represents the cross-product term, which can be positive, or negative, or zero, depending on the relationships among the projectors. Note that if the *Int* term is zero, then the above equation exactly agrees with the law of total probability. If the *Int* term is negative, however, then this can reduce the $q(Take \mid unknown)$ below both $q(Take \mid win)$ and $q(Take \mid loss)$. See Pothos and Busemeyer (2009) for more details on this application. Now we will turn to the main application for this chapter, which presents an analysis of a related puzzling finding.

11.3 Example Application to Violations of Dynamic Consistency

Multistage decisions require planning for the future across sequences of actions and events. Optimal strategies for planning multistage decisions rely on backward induction procedures, which requires the current decision to rely on plans for future decisions. Dynamic consistency requires the decision-maker to actually carry out the planned action once that decision is realized (Machina, 1989; Sarin & Wakker, 1998). This is important for optimality because if the plan is not carried out, then the optimality of the earlier decisions is compromised.

Barkan and Busemeyer (1999, 2003) experimentally investigated dynamic consistency in behavioral decision-making using a modification of the two-stage gambling task (see box 11.1).

Box 11.1
Illustration of the two-stage gambling task.

> Consider the gamble, win \$200 or lose \$100 with equal chance.
>
> During the first stage, you will be forced to play this gamble, but the outcome will not be shown until stage 2. During stage 2, you will be informed about the outcome of the first stage, and then be given a choice to play or not play the gamble on the second stage.
>
> First stage planning:
> If you win the first stage, do you plan to play the second stage? Yes or no?
> If you lose the first stage, do you plan to play the second stage? Yes or no?
>
> Second stage: Congratulations! You won \$200 in the first stage.
> Do you wish to play the gamble again a second time? Yes or no?

Participants were required to participate in a gamble (with equal chance to win x or lose y) in the first stage, and then they were asked to make a plan and give a final decision about the same gamble in a second stage. The planned decision was made before learning the outcome of the first-stage gamble ("Do you plan to play again if you win/lose the first time?"). The final decision was made after learning the outcome of the first stage (e.g., "After actually winning the first time, do you now want to play again?"). Money was paid at the end of the experiment by randomly sampling either the plan or the final decision to determine the payment.

Table 11.1, from Barkan and Busemeyer (1999), shows the basic results. The first pair of columns, labeled "Gamble," contains the payoff of a gamble; the next pair of columns, labeled "Plan assuming Win" and "Plan assuming Loss," shows the proportion of participants who took the second-stage gamble when planning assuming either a win or a loss on the first stage; the third pair of columns, labeled "Final known Win" and "Final known Loss," shows the proportion that took the gamble during the second stage after learning about winning or losing the first-stage gamble.

As can be seen, participants were dynamically inconsistent and changed their plans: following a win, participants tended to change from planning to take toward actually rejecting the gamble in the final stage; following a loss, participants tended to change from planning to reject toward actually taking the gamble in the final stage. There was very little difference between wins and losses during planning, but the differences became much larger

Table 11.1
Summary of results from the two-stage gambling task experiment of Barkan and Busemeyer (1999).

Gamble (dollars)		Choice Proportions			
Win	Loss	Plan assuming Win	Plan assuming Loss	Final known Win	Final known Loss
0.80	1	.25	.26	.20	.35
0.80	0.40	.76	.72	.69	.73
2	1	.68	.68	.60	.75
2	0.40	.84	.86	.76	.89

when the outcome was actually revealed. These findings were replicated and extended in a much-larger study by Barkan and Busemeyer (2003), which included plans and final decisions for thirty-three gambles formed from seventeen payoff conditions from 100 participants.

Barkan and Busemeyer (2003) originally developed a version of prospect theory to account for their results. This model used different utility functions for the plan and final stages, which were produced by changing the reference point of the utility function. Recently, Busemeyer et al. (2015) developed a quantum model to account for the same results, and they compared the quantum model to the earlier prospect theory model using a Bayesian model comparison procedure. Next, we describe the details of the quantum model and summarize the results of the Bayesian model comparison (see Busemeyer et al., 2015, for more detail).

11.4 Reference Point Change Model of the Two-Stage Gambling Task

The reference point model accounts for the dynamic inconsistency by shifting the reference point of a prospect theory type of utility function. The reference point of the utility function changes depending on whether or not the outcome of the first stage is experienced. Following Tversky and Shafir (1992), the utility function for a payoff value x is defined by a power function:

$$u(x) = x^{\alpha} \text{ for } x > 0 \text{ and } u(x) = -\beta \cdot |x|^{\alpha} \text{ for } x < 0.$$

The parameter $\alpha < 1$ produces risk aversion, and the parameter $\beta > 1$ produces loss aversion.

First, consider the utility function used during the planning stage before the outcome of the first stage is experienced. In the reference point model (see, e.g., Tversky & Shafir, 1992), it is assumed that the unknown outcome of the first-stage gamble is ignored, and the player decides to take the second gamble or not simply on the basis of the expected utility of the second-stage payoffs. Suppose that the gamble gives an equal chance to win ($x_W > 0$) or lose ($-x_L < 0$). Assuming equally likely events, and ignoring the first stage, the expected utility of playing the second-stage gamble during the planning stage equals

$$D_U = (.50) \cdot u(x_W) + (.50) \cdot u(-x_L) - u(0). \tag{1}$$

Next, consider the reference point model after the outcome of the first stage is experienced. If a win occurs on the first stage and the player takes the second gamble, then the possible payoff for the second stage is $(x_W + x_W)$ if a win occurs in the second stage, and $(x_W - x_L)$ if a loss occurs occurs in the second stage. If a win occurs and the player does not take the gamble, then the payoff is simply x_W from the first stage. Assuming that the events are equally likely, the expected utility difference for deciding whether to take the second-stage gamble equals

$$D_W = (.50) \cdot u(x_W + x_W) + (.50) \cdot u(x_W - x_L) - u(x_W). \tag{2}$$

If a loss occurs on the first stage, then the utility difference equals

$$D_L = (.50) \cdot u(-x_L + x_W) + (.50) \cdot u(-x_L - x_W) + u(-x_L). \tag{3}$$

To map the utility differences into choice probabilities, it is common to use a logistic choice model. The probability to take the second-stage gamble, given the information condition i (i = unknown first stage, win first stage, lose first stage), equals

$$p(\text{Take} \mid i) = 1/[1 + \exp(-\gamma \cdot D_i)],$$

and γ is a choice consistency parameter. All together, this model involves three parameters (α, β, and γ).

11.5 Quantum Model of the Two-Stage Gambling Task

Unlike the reference point model, the quantum model uses a common utility function for plans and final decisions. Instead, dynamic consistency arises from a state of uncertainty that exists during the planning process, which is resolved during the final choices.

The two-stage game involves a set of four combinations of events: {*WT*, *WR*, *LT*, *LR*}. For example, *WT* symbolizes the combination "Win the first-stage gamble" and "Take the second-stage gamble," and *LR* represents the combination "Lose the first-stage gamble" and "Reject the second-stage gamble." In the quantum model, these four combinations are represented by a four-dimensional vector space, with each dimension corresponding to one of the four combinations. The player has a belief about each event (win or lose the first play) and a potential to take each action (take or not take the second gamble). The state of the player with respect to these beliefs and action combinations is represented by a 4×1-column state vector $\psi=[\psi_{WT}, \psi_{WR}, \psi_{LT}, \psi_{LR}]^T$ that has unit length. (The notation, X^T, represents the Hermitian transpose operation). Each coordinate, ψ_{ij}, of the vector ψ represents an amplitude assigned to one of the four combinations. The squared magnitude of an amplitude equals the probability of observing a particular combination. For example, $|\psi_{LT}|^2$ equals the probability of believing that the first gamble is a loss and then deciding to take the second gamble.

The quantum model assumes that the player undergoes two stages to play the game. The first stage is an information stage, and the second stage is an action-evaluation stage.

During the first information stage, the initial state is denoted as ψ_U and is set equal to a uniform type of prior $\psi_U=[1\ 1\ 1\ 1]^T/2$. If the player does not learn the outcome of the first play (i.e., the planning stage), then they remain in this initial state. If the player learns that the first outcome was a win, then the initial state is projected onto the plane spanned by the first two dimensions (associated with "win"). This produces the projection $\psi_W=[1\ 1\ 0\ 0]^T/\sqrt{2}$, representing a state with a certain belief in a first-stage win, but that still has an equal probability of choosing actions at this information stage. If the player learns that the first outcome was a loss, then the initial state is projected onto the plane spanned by the second two dimensions (associated with "loss"). This produces the projection $\psi_L=[0\ 0\ 1\ 1]^T/\sqrt{2}$, representing a state with a known loss but equal probability of choosing actions at this information stage.

During the second evaluation stage, the payoffs are evaluated by rotating the decision-maker's information state toward or away from taking the gamble (without changing its length). Mathematically, this rotation is represented by a 4×4 unitary matrix U that transforms the state ψ into a decision state $\psi_D=U\cdot\psi$. Later in this chapter, we detail how the unitary matrix is formed on the basis of a common utility function that is used for plans and final decisions. For now, we continue to describe the computation of the choice probabilities based on the evaluation state ψ_D.

The projector used to map the evaluation state onto the subspace for choosing the action "Take the gamble" is defined by the indicator matrix $M=diag[1, 0, 1, 0]$, which picks up the first and third coordinate associated with the action to take the second-stage gamble. Then the probability to take the gamble during the plan (when the first-stage outcome is unknown) equals

$$q(\text{Take}\,|\,\text{unknown}) = ||\, M \cdot (U \cdot \psi_U)||^2.$$

The probability to take when the first-stage outcome is a known win equals

$$q(\text{Take}\,|\,\text{win}) = ||\, M \cdot (U \cdot \psi_W)||^2,$$

and the probability to take when the first-stage outcome is a known loss equals

$$q(\text{Take}\,|\,\text{loss}) = ||\, M \cdot (U \cdot \psi_L)||^2.$$

The quantum model produces by a violation of the prediction from law of total probability as follows:

$$\begin{aligned} q(\text{Take}\,|\,\text{unknown}) &= ||\, M \cdot (U \cdot \psi_U)||^2 \\ &= ||\, M \cdot U \cdot (\psi_W + \psi_L)/\sqrt{2}\ ||^2 \\ &= ||\, M \cdot U \cdot \psi_W + M \cdot U \cdot \psi_L)||^2\ /2 \\ &= [q(\text{Take}\,|\,\text{win}) + q(\text{Take}\,|\,\text{loss})]/2 + Int, \end{aligned} \tag{4}$$

where *Int* equals the cross-product

$$Int = \psi^T{}_L \cdot U^{\mathrm{T}} \cdot M \cdot U \cdot \psi_W + \psi^T{}_W \cdot U^{\mathrm{T}} \cdot M \cdot U \cdot \psi_L, \tag{5}$$

which is called the *interference term*. The term in brackets in equation (4), $[q(\text{Take}\,|\,\text{win}) + q(\text{Take}\,|\,\text{loss})]/2$, represents the total probability to take the gamble summed across the known win and known loss events. The second interference term shown in equation (5) can be positive, negative, or zero. If it is zero, then the quantum model satisfies the law of total probability; but if it is negative, then it deviates from that expected by the law of total probability.

We now describe how we construct the 4×4 unitary rotation matrix U from the utilities of the gamble. We use a construction that was successful in previous applications in other types of decision problems involving beliefs and actions (Pothos & Busemeyer, 2009; Wang & Busemeyer, 2016).

The utility differences described in equations (2) and (3) are used to construct a 4×4 Hermitian matrix:

$$H_1 = \begin{bmatrix} H_W & 0 \\ 0 & H_L \end{bmatrix}$$

$$H_W = \frac{1}{\sqrt{1+h^2{}_W}} \begin{bmatrix} h_W & 1 \\ 1 & -h_W \end{bmatrix}$$

$$H_L = \frac{1}{\sqrt{1+h^2{}_L}} \begin{bmatrix} h_L & 1 \\ 1 & -h_L \end{bmatrix},$$

with h_i defined as $h_i = tanh(D_i)$; this hyperbolic tangent transformation is needed to map utility differences onto a –1 to 1 range. The matrix H_W is used to rotate the state toward or away from taking the second-stage gamble as a function of the utility difference D_W following a win in the first stage; likewise, the matrix H_L is used to rotate the state toward or away from taking the second-stage gamble as a function of the utility difference D_L following a loss in the first stage. It is important to note that the same utility function, represented by equations (2) and (3) is used for both plans and final decisions.

Another 4×4 Hermitian matrix is also used in the quantum model:

$$H_2 = \frac{-\gamma}{\sqrt{2}} \begin{bmatrix} 1 & 0 & 1 & 0 \\ 0 & -1 & 0 & 1 \\ 1 & 0 & -1 & 0 \\ 0 & 1 & 0 & 1 \end{bmatrix}.$$

The inclusion of this second Hermitian matrix is based on previous work modeling the interactions between beliefs and actions (Pothos & Busemeyer, 2009; Wang & Busemeyer, 2016). The purpose of H_2 is to coordinate beliefs about the first-stage outcome with actions to take in the second stage. For example, if coefficient γ is positive, then H_2 rotates beliefs about winning the first stage to match taking the gamble in the second stage.

The two Hermitian matrices defined here are then used to define the unitary matrix as follows:

$$U = exp(-i \cdot (\pi 2)(H_1 + H_2)).$$

The function *exp* in this equation is a matrix exponential function (see the appendix at the end of this chapter for details). These functions are available in mathematical programming languages such as MATLAB and Mathematica. The complex number i defined by $i^2 = -1$ is needed to guarantee that U is unitary. All unitary matrices can be constructed from a Hermitian matrix by using the matrix exponential function (cf., Nielsen & Chuang, 2000 chapter 2).

All together, the quantum model entails three parameters (α, β, γ). The first two serve the same purpose as the reference point model, but the third is a uniquely quantum parameter. In summary, the two models differ in how they conceptually explain dynamic inconsistency: The reference point model uses inconsistent utility functions for the plan (equation 1) and the final (equations 2,3) decisions to produce dynamic inconsistency. According to the quantum model, inconsistency arises from a state of uncertainty that exists during the planning process, ψ_U, which is resolved during the final choices after learning the outcome of the first stage (to ψ_W or ψ_L).

11.6 Bayesian Model Comparison

The two models were compared at the individual level of analysis by evaluating the model predictions for combinations of plan and final choices observed from thirty-three gambles that each person completed. For a single person, we obtained a vector $X_i(t) = [x_{TT}(t), x_{TR}(t), x_{RT}(t), x_{RR}(t)]$, defined by $x_{jk}(t) = 1$ if event (j, k) occurs and otherwise zero, where, for example, TR is the event "Planned to take the gamble but finally rejected the gamble." To allow possible dependencies between plans and final choices within the same trial, we used an additional memory recall parameter that allows the person to simply recall and repeat the planned choice during the final choice with probability $0 \leq m \leq 1$, and otherwise make an independent choice. After including the memory recall parameter, the prediction for each trial equals

$$p_{TT} = p(T \mid \text{plan}) \cdot [(m \cdot 1) + (1 - m) \cdot p(T \mid \text{final})]$$

$$p_{TR} = p(T \mid \text{plan}) \cdot (1 - m) \cdot p(R \mid \text{final})$$

$$p_{RT} = p(R \mid \text{plan}) \cdot (1 - m) \cdot p(T \mid \text{final})$$

$$p_{RR} = p(R \mid \text{plan}) \cdot [(m \cdot 1) + (1 - m) \cdot p(R \mid \text{final})].$$

Using these model predictions, the log likelihood for a single trial equals

$$ln\ \mathrm{L}(t) = x_{TT}(t) \cdot ln\ (p_{TT}(t)) + x_{TR}(t) \cdot ln\ (p_{TR}(t)) + x_{RT}(t) \cdot ln\ (p_{RT}(t)) + x_{RR}(t) \cdot ln\ (p_{RR}(t)).$$

The data vector for a single person across all thirty-three gambles is denoted as X_i, and assuming independence across trials, the log likelihood for a single person, denoted as $ln\ L(X_i)$ for participant i, is the sum of the log likelihood for the thirty-three trials for the person.

The log likelihood for each person depends on the model (R = reference point, Q = quantum) and the vector of four parameters $\theta = (\alpha, \beta, \gamma, m)$ for each model. A Bayesian model comparison is based on the weighted average likelihood across all parameter vectors with a weight, $w_M(\theta)$ assigned to each parameter vector. We used approximately normal distributions for weights illustrated in figure 11.2.

This allowed the risk parameter α to range from strong risk aversion to strong risk seeking, with the mean located at .90 (moderate risk aversion); the loss aversion parameter ranged from less to greater than 1 with the mean located at 1.5 (moderate loss aversion), and the memory parameter ranged from perfect memory to no memory, with a mean at .50. The γ-parameter for the reference point model ranged from random choice to almost deterministic choice, with a mean at 1.5 representing a typical S-shaped choice probability function.

The Bayesian model comparison is based on a Bayes factor, which is defined as follows for each person *i:*

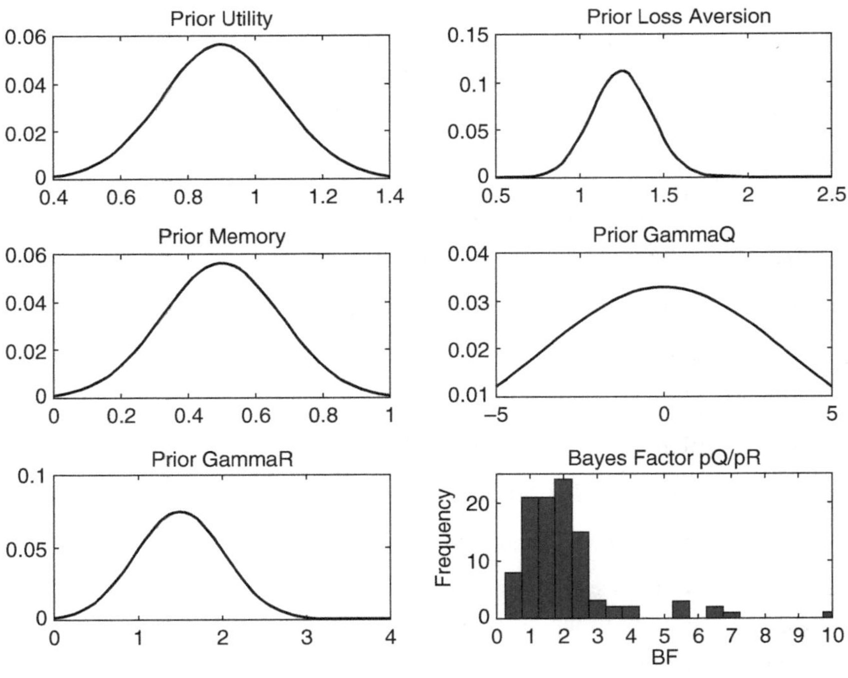

Figure 11.2
Prior distributions used for model parameters and distribution of Bayes factors from 100 participants.

$$p_M(X_i) = \sum_{\theta} w_M(\theta) \cdot \ln L_M(X_i \mid \theta)$$

$$BF_i = \frac{p_Q(X_i)}{p_R(X_i)}.$$

Figure 11.2 shows the distribution of Bayes factors across the 100 participants, and 93 percent of the participants produced Bayes factors favoring the quantum model over the reference point model. When pooling across individuals, the log Bayes factor, *ln* BF_i is more convenient to use. The total log Bayes factor, denoted *ln BF*, is obtained by summing the log Bayes factor across all 100 participants. Using the normal prior distribution, we obtained *ln* $BF = 83.05$, favoring the quantum model.

11.7 Concluding Comments

The Bayesian model comparison favored the quantum model over the reference point model to account for the dynamic inconsistency findings of Barkan and Busemeyer (2003). However, the main point of this example application is neither to claim that quantum models are uniformly better than standard decision models based on prospect theory, nor that quantum models provide the only explanation for dynamic inconsistency (see, e.g., Busemeyer et al., 2000, for another view). After all, the findings reported here and by Busemeyer et al. (2015) are specific to the Barkan and Busemeyer (2003) experiment, and specific to the quantum and reference point models that we compared (including their priors). The main point is that quantum models provide a viable new approach, and they are capable of competing with standard choice models using what may be considered the most rigorous methods for model comparison. Stronger empirical support for the quantum model described here and in Busemeyer et al. (2015) is garnered by applying the same general principles to very different types of decision problems. Essentially, the same quantum model described here for the two-stage gambling task has been used to account for puzzling findings from completely different decision paradigms, including violations of the sure thing principle found in the prisoner's dilemma game (Pothos & Busemeyer, 2009) and interference effects found with a categorization–decision-making task (Wang & Busemeyer, 2016).

This chapter provides only a brief introduction to quantum models of choice. In particular, it focused primarily on the static parts of quantum choice models. However, quantum models have also been used to describe evolution of confidence and preference across time (Kvam, Pleskac, & Busemeyer, 2015), and quantum models provide predictions for not only choice probability, but also the distribution of decision times (Busemeyer, Wang, & Townsend, 2006; Fuss & Navarro, 2013).

11.A Appendix

The matrix exponential function is defined by using the spectral decomposition of a Hermitian matrix. Suppose that H is a Hermitian matrix (i.e., a matrix that satisfies $H^T = H$). Every Hermitian matrix can be decomposed as follows:

$$H = V \cdot \Lambda \cdot V^T,$$

where V is an orthonormal eigenvector matrix (i.e., a matrix that satisfies $V^T \cdot V = I$, where I is an identity matrix); $\Lambda = diag(\ldots, \lambda_i, \ldots)$ is a diagonal matrix with real diagonal eigenvalues. The matrix exponential of H is defined as

$$exp(H) = V \cdot exp(\Lambda) \cdot V^T$$

and

$$exp(\Lambda) = \text{diag}(\ldots, \exp(\lambda_i), \ldots),$$

and $exp(\lambda_i)$ is the standard exponential function for real numbers.

12 Future Directions

The previous chapters of this book have demonstrated how large and varied and sophisticated the field of cognitive choice modeling has become. But where is it going? What will the next big steps forward be in the future? Which directions are likely to be dead ends, and which directions have good potential for growth? We speculate on these questions in this final chapter.

Deterministic utility theories (chapter 2), which morph expected utility theory into new and complicated forms, seem to be in trouble. Puzzling behavioral findings have pushed these theories farther and farther away, forcing them to relax more and more axioms, which allow increasingly complex functional forms and an alarming rise in free parameters. The difference between axiomatic and heuristic models is shrinking, and the concept of rationality is becoming controversial. Something is going wrong here. Deterministic theorists tend to shun processing mechanisms because these mechanisms seem ad hoc and add complexity to axiomatic theories. However, we speculate that this lack of inclusion of processing mechanisms may be the essential problem with evaluating utility theories. A better understanding of the dynamic and stochastic processes that underlie choice, which relate choice behavior to utility, could help save utility theory. Preference reversals and context effects that seem to violate utility theory and shape it into more complex forms may not be needed if the processing mechanisms that map utility to behavior are identified first.

Random utility models (chapter 3) provide major improvements compared to deterministic models when the goal is to quantitatively predict human choice behavior, because these models provide a way to account for variations in preference across individual differences, as well as variations in preference across replications within the same person. These models have served well as the workhorse for choice modeling by consumer and marketing researchers for many years. However, there are three main problems with most of these models: The first is that they fail to account for much of the contextual nature of human choice behavior; the second is that they provide no mechanisms for describing how long it takes to make a choice, as well as the effect of the amount of decision time on the choice that is made; and the third is that they fail to consider preference learning from experience. The sequential sampling models of preference (chapter 8) and the reinforcement learning models (chapter 9) provide natural, dynamic extensions of static random utility models, and we expect to see more applications of both of these theoretical areas to consumer and marketing research in the future.

Researchers in perception and categorization are likely to become involved in greater integration of signal detection theories (chapter 5) with sequential sampling models (chapter 6). Multidimensional signal detection theories provide a rich description of the choice stimuli. This rich description of stimuli is missing in a good deal of the modeling with random walk and diffusion models because the latter tend to simply estimate a new mean drift rate coefficient for each stimulus in an atheoretical manner. However, signal detection theory needs the dynamic processing mechanisms of sequential sampling models in order to account for decision time and speed–accuracy trade-offs. Sequential sampling models of inference provide a natural way to extend signal detection theory to model decision time and speed–accuracy trade-offs.

Although sequential sampling models (chapter 6) provide the framework for describing the dynamic evolution and accumulation of evidence, much of the previous research in this area has relied on overly simplistic assumptions concerning the dynamics of the accumulation process. In particular, much of it relies on the use of a single constant mean input drift rate that drives the accumulation process across time, as well as a constant diffusion rate that is time invariant. More realistic models are needed that allow changes in the mean drift rate and the diffusion rate across time in response to changes in attention or in the input from the environment across time during the decision process. Also, these models usually assume a constant threshold, but when considering cost of information, it becomes necessary to include dynamically changing boundaries to optimize decisions (e.g., decreasing threshold bounds with increasing cost of sampling).

There are now a large number and variety of dynamic models of preferential choice (chapters 7 and 8) that have been used to account for context effects on preference. The ability to account for a wide range of context effects is a great advantage of these models over simpler random utility models (chapter 3). The problem, though, is that the dynamic preference models are much more computationally intense to implement, fit to data, and use to estimate parameters than are the simpler random utility models. Much more theoretical and computational work is needed to make the sequential sampling models of preference more practical for use in consumer modeling and marketing research problems.

Reinforcement learning models provide a great advantage over sequential sampling choice models by describing how preferences are learned from across various experiences. The dynamic preference models describe the accumulation of preference across a time scale within a single decision, and reinforcement-learning models describe changes in the dynamics on a time scale between different decisions. However, the fundamental choice rule used in most reinforcement learning models is overly simplistic and suffers all the problems of the simple random utility models. In particular, the ratio of strength choice rule often used in reinforcement learning models fails to capture the contextual nature of choice.

Another problem is that the choice process used by reinforcement models within a single decision is static rather than dynamic, which provides no mechanism to account for decision time within a single decision. We think that it would be highly effective to integrate the dynamic models of preference within a single decision with the reinforcement learning processes across different decisions. On the one hand, this would provide a learning mechanism for sequential sampling models of preference; on the other hand, it would offer a mechanism for predicting decision time with reinforcement learning.

Recently, neuroscientists have been interested in the field of cognitive choice modeling. As pointed out in chapter 10, there are many valid reasons for understanding the neural basis of choice and decision, including to provide input about neural activation into the behavioral choice model to improve trial-by-trial prediction of behavioral choice. Alternatively, the dynamic choice model could be used to predict not only choice behavior, but also neural recording data. Often, the inclusion of brain measurements can help distinguish choice models that are almost indistinguishable at the behavioral level. However, one danger of reducing cognitive choice theory to the neural modeling level is that descriptions at this more biological level become very fine-grained and computationally intense. This could make neural models difficult or even impossible to scale up to more complicated (but more interesting and realistic) choice problems.

Quantum choice models (chapter 11) offer a completely new way of thinking about probabilistic choice. Random utility models are really deterministic at their core—a choice is determined by the utilities that exist at the moment when the choice is made. In contrast, quantum models assume that the decision-maker has an indefinite and diffusive (superposition) state immediately before the decision, and a definite state comes into existence only after a choice is probabilistically selected. Quantum models may capture the constructive nature of choice better than random utility models, and they also give a better account of sequential effects on the preference state. These models have proved successful at accounting for puzzling behavior such as order effects on preference and violations of predictions based on the classical law of total probability. In addition, quantum models have been extended to account for the dynamic evolution of preferences, and they are capable of predicting both choice and decision time. However, they have a long way to go before they can compete with the more powerful classical sequential sampling models (chapters 6–9) in terms of accounting for context effects and the distribution of choice time.

The parameters of a cognitive model are important for providing insightful measures of the latent cognitive components that combine to produce the complex behavior that is observed. Statistical methods for estimating these model parameters (chapter 4) changed greatly during the development of cognitive models. Early in the history of cognitive modeling, researchers were satisfied with parameter estimates obtained from averages across people using simple least-squares methods. Later, cognitive modelers used maximum likelihood methods to estimate parameters of individuals. However, the latter method required collecting a large amount of data at the individual level. Most recently, cognitive modelers are using hierarchical Bayesian methods to estimate model parameters. This hierarchical approach provides a higher-level model of the distribution of the parameters across individuals.

The Bayesian estimation method begins with a diffuse or uninformative prior distribution over parameters, and then it uses the data from all participants to provide a posterior distribution over the parameters rather than a single maximum likelihood point estimate for each person. However, it is important to note that the hierarchical model can have a huge effect on the posterior distribution of the model parameter estimates. Choosing an inappropriate hierarchical model (e.g., assuming a single common population when in fact there are heterogeneous groups) can produce meaningless estimates of the model parameters.

Finally, a great deal of progress has been made toward improving methods of model comparison. Early in the history of cognitive modeling, researchers simply compared goodness-of-fit statistics, such as R^2 (the proportion of variance predicted by a model), based on least-squares fits to average data. However, this method fails to account for differences in the number of model parameters (one exception is the use of a statistic called *adjusted* R^2). Later, researchers used model comparison methods based on log likelihood statistics such as the Akaike information criterion (AIC) or the Bayesian information criterion (BIC), which penalized models based on the number of parameters. However, counting the number of parameters is an inadequate way to measure model complexity and flexibility.

Most recently, cognitive modelers have used Bayes factors to compare models, which not only addresses the problem of differences in the number of parameters, but also considers the complexity of models even for those that have the same number of parameters. The Bayes factor method requires obtaining a meaningful prior distribution and computes the average log likelihood over the prior distribution. Unlike parameter estimation, which is often insensitive to the prior with sufficient data, the prior has a large effect on the Bayes factor. Therefore, we think that future researchers should conduct a sequence of experiments in which an initial experiment is used to empirically determine a meaningful prior that is then used in another experiment in a generalization test to compute the Bayes factor (see, e.g., Kary, Taylor, & Donkin, 2016).

We hope that our presentation of the field of cognitive choice modeling in this book will get researchers from both inside and outside cognitive science excited about working in this rapidly growing field.

Notes

Chapter 7

1. Some authors have renamed this topic as "value-based decision making" rather than "preferential choice," but the terms are the same.

2. This name was chosen to refer to the psychological roots based on approach-avoidance conceptions that were a part of Kurt Levin's more general dynamic theory of psychology called *field theory*.

3. Of course, this means that $w(S_i) \cdot h$ sums to h, and h can be interpreted as the probability that any state is considered during a short time step.

4. See Huang, Sen, and Szidarovszky (2012) for a more rigorous discussion of the issues involved in the convergence of the discrete time version of DFT to the OU process.

5. Also, see https://github.com/jbusemey/ChoiceModeling.git for programs that can be used to compute choice probability and distribution of stopping time for the binary choice version of DFT.

6. Also, see Huang et al. (2012) for a more rigorous discussion of the issues involved in the convergence of the discrete time version of DFT to the OU process.

7. The program used to generate these results can be downloaded from the website https://github.com/jbusemey/ChoiceModeling.git, and it can be used to fit MDFT to data.

Chapter 8

1. However, Diederich's (2003) multistage model could be used to obtain mathematically derived predictions.

References

Aerts, D. (2009). Quantum structure in cognition. *Journal of Mathematical Psychology, 53*(5), 314–348.

Ahn, W. Y., Busemeyer, J. R., Wagenmakers, E. J., & Stout, J. C. (2008). Comparison of decision learning models using the generalization criterion method. *Cognitive Science, 32*(8), 1376–1402.

Ahn, W. Y., Krawitz, A., Kim, W., Busemeyer, J. R., & Brown, J. W. (2011). A model-based fMRI analysis with hierarchical Bayesian parameter estimation. *Journal of Neuroscience, Psychology, and Economics, 4*(2), 95–110.

Alexander, G. E., DeLong, M. R., & Strick, P. L. (1986). Parallel organization of functionally segregated circuits linking basal ganglia and cortex. *Annual Review of Neuroscience, 9,* 357–381.

Allais, M. (1990). Allais paradox. In J. Eatwell, M. Milgate, & P. Newman (Eds.), *Utility and probability* (pp. 3–9). Palgrave Macmillan UK.

Anderson, N. H. (2014). *A functional theory of cognition*. Psychology Press.

Andraszewicz, S., Rieskamp, J., & Scheibehenne, B. (2015). How outcome dependencies affect decisions under risk. *Decision, 2*(2), 127–144.

Ashby, F. G. (2000). A stochastic version of general recognition theory. *Journal of Mathematical Psychology, 44*(2), 310–329.

Ashby, F. G., & Helie, S. (2011). A tutorial on computational cognitive neuroscience: modeling the neurodynamics of cognition. *Journal of Mathematical Psychology, 55*(4), 273–289.

Ashby, F. G., & Maddox, W. T. (2005). Human category learning. *Annual Review of Psychology, 56,* 149–178.

Ashby, F. G., & Soto, F. A. (2015). Multidimensional signal detection theory. In J. R. Busemeyer, Z. Wang, J. T. Townsend, & A. Eidels (Eds.), *Oxford handbook of computational and mathematical psychology* (pp. 13–34). Oxford University Press.

Ashby, F. G., & Townsend, J. T. (1986). Varieties of perceptual independence. *Psychological Review, 93*(2), 154–179.

Ashby, F. G., Alfonso-Reese, L. A., Turken, U., & Waldron, E. M. (1998). A neuropsychological theory of multiple systems in category learning. *Psychological Review, 105*(3), 442–481.

Ashby, N. J., Dickert, S., & Glöckner, A. (2012). Focusing on what you own: Biased information uptake due to ownership. *Judgment and Decision Making, 7*(3), 254–267.

Balakrishnan, J. D. (1998). Some more sensitive measures of sensitivity and response bias. *Psychological Methods, 3*(1), 68–90.

Barkan, R., & Busemeyer, J. R. (1999). Changing plans: Dynamic inconsistency and the effect of experience on the reference point. *Psychonomic Bulletin & Review, 6*(4), 547–554.

Barkan, R., & Busemeyer, J. R. (2003). Modeling dynamic inconsistency with a changing reference point. *Journal of Behavioral Decision Making, 16*(4), 235–255.

Barron, G., & Erev, I. (2003). Small feedback-based decisions and their limited correspondence to description-based decisions. *Journal of Behavioral Decision Making, 16*(3), 215–233.

Baumgartner, T., Knoch, D., Hotz, P., Eisenegger, C., & Fehr, E. (2011). Dorsolateral and ventromedial prefrontal cortex orchestrate normative choice. *Nature Neuroscience, 14*(11), 1468–1474.

Bayer, H. M., & Glimcher, P. W. (2005). Midbrain dopamine neurons encode a quantitative reward prediction error signal. *Neuron, 47*(1), 129–141.

Bechara, A., Damasio, A. R., Damasio, H., & Anderson, S. W. (1994). Insensitivity to future consequences following damage to human prefrontal cortex. *Cognition, 50*(1), 7–15.

Becker, G. M., DeGroot, M. H., & Marschak, J. (1963). Stochastic models of choice behavior. *Behavioral Science, 8*(1), 41–55.

Ben-Akiva, M., & Lerman, S. R. (2017). *Discrete choice analysis: Theory and application to travel demand.* MIT Press.

Bergner, A. S., Oppenheimer, D. M., & Detre, G. (2019). VAMP (Voting Agent Model of Preferences): A computational model of individual multi-attribute choice. *Cognition, 192*, 103971.

Berkowitsch, N. A., Scheibehenne, B., & Rieskamp, J. (2014). Rigorously testing multialternative decision field theory against random utility models. *Journal of Experimental Psychology: General, 143*(3), 1331–1348.

Berry, D. A., & Fristedt, B. (1985). *Bandit problems: Sequential allocation of experiments.* Monographs on statistics and applied probability, vol. 12. Springer.

Bhatia, S. (2013). Associations and the accumulation of preference. *Psychological Review, 120*(3), 522–543.

Bhatia, S. (2014). Sequential sampling and paradoxes of risky choice. *Psychonomic Bulletin & Review, 21*(5), 1095–1111.

Bhatia, S. (2017). Comparing theories of reference-dependent choice. *Journal of Experimental Psychology: Learning, Memory, and Cognition, 43*(9), 1490–1507.

Bhatia, S., & Mullett, T. L. (2016). The dynamics of deferred decision. *Cognitive Psychology, 86*, 112–151.

Bhattacharya, R. N., & Waymire, E. C. (1990). *Stochastic processes with applications.* Wiley.

Bhui, R., & Gershman, S. J. (2018). Decision by sampling implements efficient coding of psychoeconomic functions. *Psychological Review, 125*, 985–1001. doi:10.1037/rev0000123.

Birnbaum, M. H. (2008a). Evaluation of the priority heuristic as a descriptive model of risky decision making: Comment on Brandstätter, Gigerenzer, and Hertwig (2006). *Psychological Review, 115*(1), 253–260.

Birnbaum, M. H. (2008b). New paradoxes of risky decision making. *Psychological Review, 115*(2), 463–501.

Birnbaum, M. H., & Stegner, S. E. (1979). Source credibility in social judgment: Bias, expertise, and the judge's point of view. *Journal of Personality and Social Psychology, 37*(1), 48–74.

Birnbaum, M. H., & Zimmermann, J. M. (1998). Buying and selling prices of investments: Configural weight model of interactions predicts violations of joint independence. *Organizational Behavior and Human Decision Processes, 74*(2), 145–187.

Blavatskyy, P. R. (2007). Stochastic expected utility theory. *Journal of Risk and Uncertainty, 34*(3), 259–286.

Böckenholt, U. (2006). Thurstonian-based analyses: Past, present, and future utilities. *Psychometrika, 71*(4), 615–629.

Bogacz, R., Brown, E., Moehlis, J., Holmes, P., & Cohen, J. D. (2006). The physics of optimal decision making: a formal analysis of models of performance in two-alternative forced-choice tasks. *Psychological Review, 113*(4), 700.

Bostic, R., Herrnstein, R. J., & Luce, R. D. (1990). The effect on the preference-reversal phenomenon of using choice indifferences. *Journal of Economic Behavior & Organization, 13*(2), 193–212.

Brandenburger, A. (2014). *The language of game theory: Putting epistemics into the language of games*. World Scientific Series in Economic Science. World Scientific.

Brandstätter, E., Gigerenzer, G., & Hertwig, R. (2006). The priority heuristic: Making choices without trade-offs. *Psychological Review, 113*(2), 409–432.

Brown, S. D., & Heathcote, A. (2008). The simplest complete model of choice response time: Linear ballistic accumulation. *Cognitive Psychology, 57*(3), 153–178.

Browne, M. W. (2000). Cross-validation methods. *Journal of Mathematical Psychology, 44*(1), 108–132.

Bruza, P. D., Wang, Z., & Busemeyer, J. R. (2015). Quantum cognition: A new theoretical approach to psychology. *Trends in Cognitive Sciences, 19*(7), 383–393.

Busemeyer, J. R. (1985). Decision making under uncertainty: A comparison of simple scalability, fixed-sample, and sequential-sampling models. *Journal of Experimental Psychology: Learning, Memory, and Cognition, 11*(3), 538–564.

Busemeyer, J. R., & Bruza, P. D. (2012). *Quantum models of cognition and decision*. Cambridge University Press.

Busemeyer, J. R., & Diederich, A. (2002). Survey of decision field theory. *Mathematical Social Sciences, 43*(3), 345–370.

Busemeyer, J. R., & Diederich, A. (2010). *Cognitive modeling*. SAGE.

Busemeyer, J. R., & Myung, I. J. (1992). An adaptive approach to human decision making: Learning theory, decision theory, and human performance. *Journal of Experimental Psychology: General, 121*(2), 177–194.

Busemeyer, J. R., & Townsend, J. T. (1992). Fundamental derivations from decision field theory. *Mathematical Social Sciences, 23*(3), 255–282.

Busemeyer, J. R., & Townsend, J. T. (1993). Decision field theory: A dynamic-cognitive approach to decision making in an uncertain environment. *Psychological Review, 100*(3), 432–459.

Busemeyer, J. R., & Wang, Y. M. (2000). Model comparisons and model selections based on generalization criterion methodology. *Journal of Mathematical Psychology, 44*(1), 171–189.

Busemeyer, J. R. Gluth, S., Rieskamp, J., & Turner, B. (2019). Cognitive and neural bases of multi-attribute, multi-alternative, value-based decisions. *Trends in Cognitive Sciences, 23*(3), 251–263.

Busemeyer, J. R., Matthew, M. R., & Wang, Z. (2006). A quantum information processing explanation of disjunction effects. In R. Sun & N. Miyake (Eds.), *The 28th Annual Conference of the Cognitive Science Society & the 5th International Conference of the Cognitive Science* (pp. 131–135). Erlbaum.

Busemeyer, J. R., Townsend, J. T., & Stout, J. C. (2002). Motivational underpinnings of utility in decision making: Decision field theory analysis of deprivation and satiation. In S. Moore & M. Oaksford (Eds.), *Emotional cognition* (pp. 197–220). Amsterdam: John Benjamins.

Busemeyer, J. R., Wang, Z., & Shiffrin, R. M. (2015). Bayesian model comparison favors quantum over standard decision theory account of dynamic inconsistency. *Decision, 2*(1), 1–12.

Busemeyer, J. R., Wang, Z., & Townsend, J. T. (2006). Quantum dynamics of human decision-making. *Journal of Mathematical Psychology, 50*(3), 220–241.

Busemeyer, J. R., Wang, Z., Townsend, J. T., & Eidels, A. (2015). *Oxford handbook of computational and mathematical psychology.* Oxford University Press.

Busemeyer, J. R., Weg, E., Barkan, R., Li, X., & Ma, Z. (2000). Dynamic and consequential consistency of choices between paths of decision trees. *Journal of Experimental Psychology General, 129*(4), 530–545.

Bush, R. R., & Mosteller, F. (1955). *Stochastic models for learning.* Wiley.

Colas, J. T. (2017). Value-based decision making via sequential sampling with hierarchical competition and attentional modulation. *PloS One, 12*(10), e0186822.

Dai, J., & Busemeyer, J. R. (2014). A probabilistic, dynamic, and attribute-wise model of intertemporal choice. *Journal of Experimental Psychology: General, 143*(4), 1489–1514.

Damasio, A. R. (2006). *Descartes' error.* Random House.

Daw, N., & O'Doherty, J. (2014) Multiple systems for value learning. In P. Glimcher & E. Fehr (Eds.), *Neuroeconomics: Decision making and the brain* (2nd ed., pp. 393–410). Elsevier.

Daw, N. D., & Tobler, P. (2014) Value learning through reinforcement: The basics of dopamine and reinforcement learning. In P. Glimcher & E. Fehr (Eds.), *Neuroeconomics: Decision making and the brain* (2nd ed., pp. 283–296). Elsevier.

Daw, N. D., Niv, Y., & Dayan, P. (2005). Uncertainty-based competition between prefrontal and dorsolateral striatal systems for behavioral control. *Nature Neuroscience, 8*(12), 1704–1711.

Dayan, P., & Abbott, L. F. (2001). *Theoretical neuroscience, vol. 806.* MIT Press.

Dayan, P., & Niv, Y. (2008). Reinforcement learning: The good, the bad and the ugly. *Current Opinion in Neurobiology, 18*(2), 185–196.

DeGroot, M. H. (1970). *Optimal statistical decisions.* McGraw-Hill.

DeGroot, M. H. (1974). Reaching a consensus. *Journal of the American Statistical Association, 69*(345), 118–121.

Diederich, A. (2003a). Decision making under conflict: Decision time as a measure of conflict strength. *Psychonomic Bulletin & Review, 10*(1), 167–176.

Diederich, A. (2003b). MDFT account of decision making under time pressure. *Psychonomic Bulletin & Review, 10*(1), 157–166.

Diederich, A. (2016). A multistage attention-switching model account for payoff effects on perceptual decision tasks with manipulated processing order. *Decision, 3*(2), 81–114.

Diederich, A., & Busemeyer, J. R. (1999). Conflict and the stochastic-dominance principle of decision making. *Psychological Science, 10*(4), 353–359.

Diederich, A., & Busemeyer, J. R. (2003). Simple matrix methods for analyzing diffusion models of choice probability, choice response time, and simple response time. *Journal of Mathematical Psychology, 47*(3), 304–322.

Diederich, A., & Oswald, P. (2016). Multi-stage sequential sampling models with finite or infinite time horizon and variable boundaries. *Journal of Mathematical Psychology, 74,* 128–145.

Diederich, A., & Trueblood, J. S. (2018). A dynamic dual process model of risky decision making. *Psychological Review, 125*(2), 270–292.

Dirac, P. A. M. (1930). *The principles of quantum mechanics*. Clarendon Press.

Dorfman, D. D., & Alf, E. (1969). Maximum-likelihood estimation of parameters of signal-detection theory and determination of confidence intervals—rating-method data. *Journal of Mathematical Psychology, 6*(3), 487–496.

Dorfman, D. D., & Biderman, M. (1971). A learning model for a continuum of sensory states. *Journal of Mathematical Psychology, 8,* 264–284.

Dror, I. E., Basola, B., & Busemeyer, J. R. (1999). Decision making under time pressure: An independent test of sequential sampling models. *Memory & Cognition, 27*(4), 713–725.

Ellsberg, D. (1961). Risk, ambiguity, and the Savage axioms. *Quarterly Journal of Economics,* 643–669.

Erev, I., & Barron, G. (2005). On adaptation, maximization, and reinforcement learning among cognitive strategies. *Psychological Review, 112*(4), 912–931.

Erev, I., & Roth, A. E. (1998). Predicting how people play games: Reinforcement learning in experimental games with unique, mixed-strategy equilibria. *American Economic Review, 88*(4), 848–881.

Erev, I., & Roth, A. E. (2014). Maximization, learning, and economic behavior. *Proceedings of the National Academy of Sciences, 111*(Suppl 3), 10818–10825.

Erev, I., Ert, E., Roth, A. E., Haruvy, E., Herzog, S. M., Hau, R., ... & Lebiere, C. (2010). A choice prediction competition: Choices from experience and from description. *Journal of Behavioral Decision Making, 23*(1), 15–47.

Estes, W. K. (1950). Toward a statistical theory of learning. *Psychological Review, 57*(2), 94–107.

Estes, W. K., & Maddox, W. T. (2005). Risks of drawing inferences about cognitive processes from model fits to individual versus average performance. *Psychonomic Bulletin & Review, 12*(3), 403–408.

Evans, N. J., Holmes, W. R., & Trueblood, J. S. (2019). Response-time data provide critical constraints on dynamic models of multi-alternative, multi-attribute choice. *Psychonomic Bulletin & Review, 26*(3), 901–933.

Everitt, B. J., & Robbins, T. W. (2005). Neural systems of reinforcement for drug addiction: From actions to habits to compulsion. *Nature Neuroscience, 8*(11), 1481–1489.

Fakhari, P., Khodadadi, A., & Busemeyer, J. R. (2018). The detour problem in a stochastic environment: Tolman revisited. *Cognitive Psychology, 101,* 29–49.

Farahmand, F. (2017a). Decision and experienced utility: Computational applications in privacy decision making. *IEEE Computer, 15*(6), 68–72.

Farahmand, F. (2017b). The importance of human information processing: A behavioral economics model for predicting domain name choice. *IEEE Computer, 50*(9), 67–74.

Fehr, E., & Schmidt, K. M. (1999). A theory of fairness, competition, and cooperation. *Quarterly Journal of Economics, 114*(3), 817–868.

Feller, W. (1968). *An introduction to probability theory and its applications: Volume I.* John Wiley & Sons.

Fiorillo, C. D., Tobler, P. N., & Schultz, W. (2003). Discrete coding of reward probability and uncertainty by dopamine neurons. *Science, 299*(5614), 1898–1902.

Fontanesi, L., Gluth, S., Spektor, M. S., & Rieskamp, J. (2019). A reinforcement learning diffusion decision model for value-based decisions. *Psychonomic Bulletin & Review,* 1–23.

Fox, C. R., & Hadar, L. (2006). "Decisions from experience" = sampling error + prospect theory: Reconsidering Hertwig, Barron, Weber, & Erev (2004). *Judgment and Decision Making, 1*(2), 159–161.

Fuss, I. G., & Navarro, D. J. (2013). Open parallel cooperative and competitive decision processes: A potential provenance for quantum probability decision models. *Topics in Cognitive Science, 5*(4), 818–843.

Gao, J., & Lee, J. D. (2006). Extending the decision field theory to model operators' reliance on automation in supervisory control situations. *IEEE Transactions on Systems, Man, and Cybernetics—Part A: Systems and Humans, 36*(5), 943–959.

Gao, J., Lee, J. D., & Zhang, Y. (2006). A dynamic model of interaction between reliance on automation and cooperation in multi-operator multi-automation situations. *International Journal of Industrial Ergonomics, 36*(5), 511–526.

Gelman, A., Carlin, J. B., & Stern, H. S. (2016). *Bayesian data analysis.* CRC Press.

Glimcher, P. W. (2005). Indeterminacy in brain and behavior. *Annual Review of Psychology, 56,* 25–56.

Glimcher, P. W., & Fehr, E. (Eds.). (2014). *Neuroeconomics: Decision making and the brain.* Elsevier.

Gold, J. I., & Heekeren, H. R. (2014). Neural mechanisms for perceptual decision making. In P. Glimcher & E. Fehr (Eds.), *Neuroeconomics: Decision making and the brain* (2nd ed., pp. 356–371). Elsevier.

Gold, J. I., & Shadlen, M. N. (2002). Banburismus and the brain: Decoding the relationship between sensory stimuli, decisions, and reward. *Neuron, 36*(2), 299–308.

Gold, J. I., & Shadlen, M. N. (2007). The neural basis of decision making. *Annual Review of Neuroscience, 30,* 535–574.

Gonzalez, C., & Dutt, V. (2011). Instance-based learning: Integrating sampling and repeated decisions from experience. *Psychological Review, 118*(4), 523–551.

Gonzalez, C., Lerch, J. F., & Lebiere, C. (2003). Instance-based learning in dynamic decision making. *Cognitive Science, 27*(4), 591–635.

Gonzalez-Vallejo, C. (2002). Making trade-offs: A probabilistic and context-sensitive model of choice behavior. *Psychological Review, 109*(1), 137–155.

Grant, S., Contoreggi, C., & London, E. D. (2000). Drug abusers show impaired performance in a laboratory test of decision making. *Neuropsychologia, 38*(8), 1180–1187.

Green, D. M., & Swets, J. A. (1966). *Signal detection and psychophysics.* Wiley.

Grether, D. M., & Plott, C. R. (1979). Economic theory of choice and the preference reversal phenomenon. *American Economic Review, 69*(4), 623–638.

Grossberg, S. (1988). Nonlinear neural networks: Principles, mechanisms, and architectures. *Neural Networks, 1*(1), 17–61.

Gureckis, T. M., & Love, B. C. (2009a). Learning in noise: Dynamic decision-making in a variable environment. *Journal of Mathematical Psychology, 53*(3), 180–193.

Gureckis, T. M., & Love, B. C. (2009b). Short-term gains, long-term pains: How cues about state aid learning in dynamic environments. *Cognition, 113*(3), 293–313.

Halff, H. M. (1976). Choice theories for differentially comparable alternatives. *Journal of Mathematical Psychology, 14*(3), 244–246.

Hampton, A. N., Bossaerts, P., O'Doherty, J. P. (2006). The role of the ventromedial prefrontal cortex in abstract state-based inference during decision making in humans. *Journal of Neuroscience, 26*(32), 8360–8367.

Hancock, T. O., Hess, S., & Choudhury, C. F. (2018). Decision field theory: Improvements to current methodology and comparisons with standard choice modelling techniques. *Transportation Research Part B: Methodological, 107,* 18–40.

Hanes, D. P., & Schall, J. D. (1996). Neural control of voluntary movement initiation. *Science, 274*(5286), 427–430.

Hau, R., Pleskac, T. J., & Hertwig, R. (2010). Decisions from experience and statistical probabilities: Why they trigger different choices than a priori probabilities. *Journal of Behavioral Decision Making, 23*(1), 48–68.

Hau, R., Pleskac, T. J., Kiefer, J., & Hertwig, R. (2008). The description–experience gap in risky choice: The role of sample size and experienced probabilities. *Journal of Behavioral Decision Making, 21*(5), 493–518.

Hausman, J. A., & Wise, D. A. (1978). A conditional probit model for qualitative choice: Discrete decisions recognizing interdependence and heterogeneous preferences. *Econometrica: Journal of the Econometric Society,* 403–426.

Heekeren, H. R., Marrett, S., Bandettini, P. A., & Ungerleider, L. G. (2004). A general mechanism for perceptual decision-making in the human brain. *Nature, 431*(7010), 859–862.

Hertwig, R., & Erev, I. (2009). The description–experience gap in risky choice. *Trends in Cognitive Sciences, 13*(12), 517–523.

Hertwig, R., Barron, G., Weber, E. U., & Erev, I. (2004). Decisions from experience and the effect of rare events in risky choice. *Psychological Science, 15*(8), 534–539.

Hess, S., & Daly, A. (2014). *Handbook of choice modelling.* Edward Elgar.

Hey, J. D., & Orme, C. (1994). Investigating generalizations of expected utility theory using experimental data. *Econometrica: Journal of the Econometric Society,* 1291–1326.

Higham, D. J. (2001). An algorithmic introduction to numerical simulation of stochastic differential equations. *SIAM Review, 43*(3), 525–546.

Hogg, R. V., & Craig, A. T. (1995). *Introduction to mathematical statistics.* 5th ed. Prentice Hall.

Hotaling, J. M., & Rieskamp, J. (2018). A quantitative test of computational models of multialternative context effects. *Decision, 6* (3) 201–222.

Hotaling, J. M., Busemeyer, J. R., & Li, J. (2010). Theoretical developments in decision field theory: Comment on Tsetsos, Usher, and Chater (2010). *Psychological Review, 117*(4), 1294–1298.

Huang, K., Sen, S., & Szidarovszky, F. (2012). Connections among decision field theory models of cognition. *Journal of Mathematical Psychology, 56*(5), 287–296.

Huber, J., Payne, J. W., & Puto, C. (1982). Adding asymmetrically dominated alternatives: Violations of regularity and the similarity hypothesis. *Journal of Consumer Research, 9*(1), 90–98.

Hutchinson, J. W., Kamakura, W. A., & Lynch Jr., J. G. (2000). Unobserved heterogeneity as an alternative explanation for "reversal" effects in behavioral research. *Journal of Consumer Research, 27*(3), 324–344.

Iverson, G., & Falmagne, J. C. (1985). Statistical issues in measurement. *Mathematical Social Sciences, 10*(2), 131–153.

Jessup, R. K., Bishara, A. J., & Busemeyer, J. R. (2008). Feedback produces divergence from prospect theory in descriptive choice. *Psychological Science, 19*(10), 1015–1022.

Jessup, R. K., Busemeyer, J. R., & Brown, J. W. (2010). Error effects in anterior cingulate cortex reverse when error likelihood is high. *Journal of Neuroscience, 30,* 3467–3472.

Jessup, R. K., Veinott, E. S., Todd, P. M., & Busemeyer, J. R. (2009). Leaving the store empty-handed: Testing explanations for the too-much-choice effect using decision field theory. *Psychology & Marketing, 26*(3), 299–320.

Johnson, J. G. (2006). Cognitive modeling of decision making in sports. *Psychology of Sport and Exercise, 7*(6), 631–652.

Johnson, J. G., & Busemeyer, J. R. (2005). A dynamic, stochastic, computational model of preference reversal phenomena. *Psychological Review, 112*(4), 841–861.

Johnson, J. G., & Busemeyer, J. R. (2016). A computational model of the attention process in risky choice. *Decision, 3*(4), 254–280.

Jones, M., & Dzhafarov, E. N. (2014). Unfalsifiability and mutual translatability of major modeling schemes for choice reaction time. *Psychological Review, 121*(1), 1–32.

Juslin, P., Olsson, H., & Olsson, A.-C. (2003). Exemplar effects in categorization and multiple-cue judgment. *Journal of Experimental Psychology: General, 132,* 133–156.

Kable, J. W., & Glimcher, P. W. (2007). The neural correlates of subjective value during intertemporal choice. *Nature Neuroscience, 10*(12), 1625–1633.

Kahneman, D. (2003). Maps of bounded rationality: Psychology for behavioral economics. *American Economic Review, 93*(5), 1449–1475.

Kahneman, D., & Tversky, A. (1979). Prospect theory: An analysis of decision under risk. *Econometrica: Journal of the Econometric Society*, 263–291.

Karlin, S., & Taylor, H. E. (1981). A second course in stochastic processes. Elsevier.

Kary, A., Taylor, R., & Donkin, C. (2016). Using Bayes factors to test the predictions of models: A case study in visual working memory. *Journal of Mathematical Psychology, 72,* 210–219.

Keeney, R. L., & Raiffa, H. (1993). *Decisions with multiple objectives: preferences and value trade-offs*. Cambridge University Press.

Keren, G., & Wagenaar, W. A. (1987). Violation of utility theory in unique and repeated gambles. *Journal of Experimental Psychology: Learning, Memory, and Cognition, 13*(3), 387–391.

Khrennikov, A. Y. (2010). *Ubiquitous quantum structure.* Springer.

Kolmogorov, A. N. (1933). *Foundations of probability.* Dover

Krajbich, I., & Rangel, A. (2011). Multialternative drift-diffusion model predicts the relationship between visual fixations and choice in value-based decisions. *Proceedings of the National Academy of Sciences, 108*(33), 13852–13857.

Krajbich, I., Armel, C., & Rangel, A. (2010). Visual fixations and the computation and comparison of value in simple choice. *Nature Neuroscience, 13*(10), 1292–1298.

Krajbich, I., Hare, T., Bartling, B., Morishima, Y., & Fehr, E. (2015). A common mechanism underlying food choice and social decisions. *PLoS Computational Biology, 11*(10), e1004371.

Krajbich, I., Lu, D., Camerer, C., & Rangel, A. (2012). The attentional drift-diffusion model extends to simple purchasing decisions. *Frontiers in Psychology, 3,* 193.

Krantz, D., Luce, D., Suppes, P., & Tversky, A. (1971). *Foundations of measurement, Vol. I: Additive and polynomial representations.* Academic Press.

Krantz, D. H., & Tversky, A. (1971). Conjoint-measurement analysis of composition rules in psychology. *Psychological Review, 78*(2), 151–169.

Kruschke, J. K. (2015). *Doing Bayesian data analysis: A tutorial with R, JAGS, and Stan.* Elsevier

Kvam, P. D. (2019). A geometric framework for modeling dynamic decisions among arbitrarily many alternatives. *Journal of Mathematical Psychology, 91,* 14–37.

Kvam, P. D. & Busemeyer, J. R. (2020) A distributional and dynamic theory of pricing and preference. *Psychological Review.*

Kvam, P. D., Pleskac, T. J., Yu, S., & Busemeyer, J. R. (2015). Interference effects of choice on confidence: Quantum characteristics of evidence accumulation. *Proceedings of the National Academy of Sciences, 112*(34), 10645–10650.

Laming, D. R. J. (1968). *Information theory of choice-reaction times*. Academic Press.

Lau, B. (2015) MatlabStan: the MATLAB interface to Stan.

Lebiere, C., Stewart, T., & West, R. (2009). Applying cognitive architectures to decision-making: How cognitive theory and the equivalence measure triumphed in the technion prediction tournament. *Proceedings of the Annual Meeting of the Cognitive Science Society, 31*. Retrieved from https://escholarship.org/uc/item/4622n7tt.

Lee, M. D. (2008). Three case studies in the Bayesian analysis of cognitive models. *Psychonomic Bulletin & Review, 15*(1), 1–15.

Lee, M. D., & Wagenmakers, E. J. (2014). *Bayesian cognitive modeling: A practical course*. Cambridge University Press.

Lee, S., & Son, Y. J. (2020). Extended decision field theory with social-learning for long-term decision-making processes in social networks. *Information Sciences, 512,* 1293–1307.

Lee, S., Son, Y. J., & Jin, J. (2008). Decision field theory extensions for behavior modeling in dynamic environment using Bayesian belief network. *Information Sciences, 178*(10), 2297–2314.

Lee, S., Son, Y. J., & Jin, J. (2010). An integrated human decision making model for evacuation scenarios under a BDI framework. *ACM Transactions on Modeling and Computer Simulation (TOMACS), 20*(4), 23.

Lieder, F., & Griffiths, T. L. (2017). Strategy selection as rational metareasoning. *Psychological Review, 124*(6), 762.

Levy, D. J., & Glimcher, P. W. (2012). The root of all value: A neural common currency for choice. *Current Opinion in Neurobiology, 22*(6), 1027–1038.

Lewandowsky, S., & Farrell, S. (2011). *Computational modeling in cognition: Principles and practice*. SAGE Publications.

Lewis, R. L., Howes, A., & Singh, S. (2014). Computational rationality: Linking mechanism and behavior through bounded utility maximization. *Topics in Cognitive Science, 6,* 279–311.

Lichtenstein, S., & Slovic, P. (1971). Reversals of preference between bids and choices in gambling decisions. *Journal of Experimental Psychology, 89*(1), 46–55.

Liew, S. X., Howe, P. D., & Little, D. R. (2016). The appropriacy of averaging in the study of context effects. *Psychonomic Bulletin & Review, 23*(5), 1639–1646.

Link, S. W. (1975). The relative judgment theory of two choice response time. *Journal of Mathematical Psychology, 12*(1), 114–135.

Link, S. W. (1992). *The wave theory of difference and similarity*. Psychology Press.

Link, S. W., & Heath, R. A. (1975). A sequential theory of psychological discrimination. *Psychometrika, 40*(1), 77–105.

Liu, T., & Pleskac, T. J. (2011). Neural correlates of evidence accumulation in a perceptual decision task. *Journal of Neurophysiology, 106*(5), 2383–2398.

Loewenstein, G., & Elster, J. (Eds.). (1992). *Choice over time.* Russell Sage Foundation.

Lopes, L. L. (1987). Between hope and fear: The psychology of risk. In L. Berkowitz (Ed.), *Advances in experimental social psychology* (vol. 20, pp. 255–295). Academic Press.

Louviere, J. J., Hensher, D. A., & Swait, J. D. (2000). *Stated choice methods: Analysis and applications.* Cambridge University Press.

Luce, R. D. (1959, 2005). *Individual choice behavior: A theoretical analysis.* Courier Corporation.

Luce, R. D. (1977). The choice axiom after twenty years. *Journal of Mathematical Psychology, 15*(3), 215–233.

Luce, R. D. (1986). *Response times: Their role in inferring elementary mental organization (No. 8).* Oxford University Press on Demand.

Luce, R. D. (2000). *Utility of gains and losses: Measurement-theoretical and experimental approaches.* Lawrence Erlbaum.

Luce, R. D., & Raiffa, H. (1957). *Games and decisions: Introduction and critical survey.* Wiley.

Machina, M. J. (1989). Dynamic consistency and non-expected utility models of choice under uncertainty. *Journal of Economic Literature, 27*(4), 1622–1668.

MacKay, D. B., & Zinnes, J. L. (1986). A probabilistic model for the multidimensional scaling of proximity and preference data. *Marketing Science, 5*(4), 325–344.

Macmillan, N. A. (2002). Signal detection theory. In J. Wixted (Ed.), *Stevens' handbook of experimental psychology.* Wiley.

Macmillan, N. A., & Creelman, C. D. (2004). *Detection theory: A user's guide.* Psychology Press.

Maddox, W. T., & Ing, D, A. (2005). Delayed feedback disrupts the procedural-learning system but not the hypothesis-testing system in perceptual category learning. *Journal of Experimental Psychology: Learning, Memory, and Cognition, 31*(1), 100.

Marley, A. A. J., & Colonius, H. (1992). The "horse race" random utility model for choice probabilities and reaction times, and its compering risks interpretation. *Journal of Mathematical Psychology, 36*(1), 1–20.

McFadden, D. (1981). Economic models of probabilistic choice. In C. F. Manski & D. McFadden (Eds.), *Structural analysis of discrete data with econometric applications* (pp. 198–272). MIT Press.

Mellers, B. A., & Biagini, K. (1994). Similarity and choice. *Psychological Review, 101*(3), 505–518.

Milosavljevic, M., Malmaud, J., Huth, A., Koch, C., & Rangel, A. (2010). The drift diffusion model can account for the accuracy and reaction time of value-based choices under high and low time pressure. *Judgment and Decision Making, 5,* 437–449.

Miyamoto, J. M., & Eraker, S. A. (1988). A multiplicative model of the utility of survival duration and health quality. *Journal of Experimental Psychology: General, 117*(1), 3.

Mogiliansky, A. L., Zamir, S., & Zwirn, H. (2009). Type indeterminacy: A model of the KT (Kahneman–Tversky)-man. *Journal of Mathematical Psychology, 53*(5), 349–361.

Mohr, P. N., Heekeren, H. R., & Rieskamp, J. (2017). Attraction effect in risky choice can be explained by subjective distance between choice alternatives. *Scientific Reports,* 7.

Molloy, M. F., Galdo, M., Bahg, G., Liu, Q., & Turner, B. M. (2019). What's in a response time? On the importance of response time measures in constraining models of context effects. *Decision, 6*(2), 171–200.

Montague, P. R., Dayan, P., Person, C., & Sejnowski, T. J. (1995). Bee foraging in uncertain environments using predictive Hebbian learning. *Nature, 377,* 725–728.

Montague, P. R., Dayan, P., & Sejnowski, T. J. (1994). Foraging in an uncertain environments using predictive Hebbian learning. In J. D. Cowan, G. Tesauro, & J. Alspector (Eds.), *Advances in neural information processing systems* (Vol. 6, pp. 598–605). San Mateo.

Moran, R., Teodorescu, A. R., & Usher, M. (2015). Post choice information integration as a causal determinant of confidence: Novel data and a computational account. *Cognitive Psychology, 78,* 99–147.

Mulder, M. J., Wagenmakers, E. J., Ratcliff, R., Boekel, W., & Forstmann, B. U. (2012). Bias in the brain: A diffusion model analysis of prior probability and potential payoff. *Journal of Neuroscience, 32*(7), 2335–2343.

Navarro-Martinez, D., Loomes, G., Isoni, A., Butler, D., & Alaouli, L. (2018) Boundedly rational expected utility theory. *Journal of Risk and Uncertainty, 57,* 199–223

Newsome, W.T., Britten, K. H., & Movshon, J.A. (1989). Neuronal correlates of a perceptual decision. *Nature, 341,* 52–54.

Nielsen, M. A., & Chuang, I. (2000). *Quantum computation and quantum information.* Cambridge University Press.

Niv, Y. (2009). Reinforcement learning in the brain. *Journal of Mathematical Psychology, 53*(3), 139–154.

Noguchi, T., & Stewart, N. (2014). In the attraction, compromise, and similarity effects, alternatives are repeatedly compared in pairs on single dimensions. *Cognition, 132*(1), 44–56.

Noguchi, T., & Stewart, N. (2018). Multialternative decision by sampling: A model of decision making constrained by process data. *Psychological Review, 125,* 512–544.

Nosofsky, R. M., & Palmeri, T. J. (1997). An exemplar-based random walk model of speeded classification. *Psychological Review, 104*(2), 266–300.

O'Doherty, J. P., Dayan, P., Friston, K., Critchley, H., & Dolan, R. J. (2003). Temporal difference models and reward-related learning in the human brain. *Neuron, 38*(2), 329–337.

Osman, A., Bashore, T. R., Coles, M. G., Donchin, E., & Meyer, D. E. (1992). On the transmission of partial information: Inferences from movement-related brain potentials. *Journal of Experimental Psychology: Human Perception and Performance, 18*(1), 217–232.

Otter, T., Allenby, G. M., & Van Zandt, T. (2008). An integrated model of discrete choice and response time. *Journal of Marketing Research, 45*(5), 593–607.

Parducci, A. (1965). Category judgment: A range-frequency model. *Psychological Review, 72*(6), 407–418.

Payne, J. W., Bettman, J. R., & Johnson, E. J. (1993). *The adaptive decision maker.* Cambridge University Press.

Pedersen, M. L., Frank, M. J., & Biele, G. (2017). The drift diffusion model as the choice rule in reinforcement learning. *Psychonomic Bulletin & Review, 24*(4), 1234–1251.

Peterson, W. W., Birdsall, T. G., & Fox, W. C. (1954). The theory of signal detectability. *Transactions of the IRE Professional Group on Information Theory, PGIT-4,* 171–212.

Petrusic, W. M., & Jamieson, D. G. (1978). Relation between probability of preferential choice and time to choose changes with practice. *Journal of Experimental Psychology: Human Perception and Performance, 4*(3), 471–482.

Pettibone, J. C. (2012). Testing the effect of time pressure on asymmetric dominance and compromise decoys in choice. *Judgment and Decision Making, 7*(4), 513–523.

Philiastides, M. G., & Heekeren, H. R. (2009). Spatiotemporal characteristics of perceptual decision making in the human brain. In J. C. Dreher & L. Tremplay (Eds.), *Handbook of reward and decision making* (pp. 185–212). Academic Press.

Pleskac, T. J., & Busemeyer, J. R. (2010). Two-stage dynamic signal detection: A theory of choice, decision time, and confidence. *Psychological Review, 117*(3), 864–901.

Pleskac, T. J., & Hertwig, R. (2014). Ecologically rational choice and the structure of the environment. *Journal of Experimental Psychology: General, 143*(5), 2000–2019.

Plonsky, O., Teodorescu, K., & Erev, I. (2015). Reliance on small samples, the wavy recency effect, and similarity-based learning. *Psychological Review, 122*(4), 621–647.

Pothos, E. M., & Busemeyer, J. R. (2009). A quantum probability explanation for violations of "rational" decision theory. *Proceedings of the Royal Society of London B: Biological Sciences,* 2171–2178.

Pothos, E. M., & Busemeyer, J. R. (2013). Can quantum probability provide a new direction for cognitive modeling? *Behavioral and Brain Sciences, 36*(3), 255–274.

Purcell, B., Heitz, R., Cohen, J., Schall, J., Logan, G., & Palmeri, T. (2010). Neurally-constrained modeling of perceptual decision making. *Psychological Review, 117,* 1113–1143.

Puterman, M. L. (2014). Markov decision processes: discrete stochastic dynamic programming. Wiley.

Quiggin, J. (1982). A theory of anticipated utility. *Journal of Economic Behavior & Organization, 3*(4), 323–343.

Rangel, A., & Clithero, J. A. (2014). The computation of stimulus values in simple choice. In P. Glimcher & E. Fehr (Eds.), *Neuroeconomics: Decision making and the brain* (2nd ed., 125–147). Elsevier.

Raab, M. (2002). T-ECHO: Model of decision making to explain behaviour in experiments and simulations under time pressure. *Psychology of Sport and Exercise, 3*(2), 151–171.

Ratcliff, R. (1978). A theory of memory retrieval. *Psychological Review, 85*(2), 59–108.

Ratcliff, R. (2013). Parameter variability and distributional assumptions in the diffusion model. *Psychological Review, 120*(1), 281–292.

Ratcliff, R. (2018). Decision making on spatially continuous scales. *Psychological Review, 125*(6), 888–935.

Ratcliff, R., & Childers, R. (2015). Individual differences and fitting methods for the two-choice diffusion model of decision making. *Decision, 2*(4), 237–279.

Ratcliff, R., & Smith, P. L. (2004). A comparison of sequential sampling models for two-choice reaction time. *Psychological Review, 111*(2), 333.

Ratcliff, R., & Smith, P. (2015). Modeling simple decisions and applications using a diffusion model. In J. R. Busemeyer, Z. Wang, J. T. Townsend, & A. Eidels (Eds.), *Oxford library of psychology. The Oxford handbook of computational and mathematical psychology* (pp. 35–62). Oxford University Press.

Ratcliff, R., & Starns, J. J. (2009). Modeling confidence and response time in recognition memory. *Psychological Review, 116*(1), 59–83.

Ratcliff, R., & Starns, J. J. (2013). Modeling confidence judgments, response times, and multiple choices in decision making: Recognition memory and motion discrimination. *Psychological Review, 120*(3), 697–719.

Ratcliff, R., Hasegawa, Y. T., Hasegawa, R. P., Smith, P. L., & Segraves, M. A. (2007). Dual diffusion model for single-cell recording data from the superior colliculus in a brightness-discrimination task. *Journal of Neurophysiology, 97,* 1756–1774.

Ratcliff, R., Philiastides, M. G., & Sajda, P. (2009). Quality of evidence for perceptual decision making is indexed by trial-to-trial variability of the EEG. *Proceedings of the National Academy of Sciences of the United States, 106,* 6539–6544.

Ratcliff, R., Smith, P. L., Brown, S. D., & McKoon, G. (2016). Diffusion decision model: Current issues and history. *Trends in Cognitive Sciences, 20*(4), 260–281.

Regenwetter, M., Dana, J., & Davis-Stober, C. P. (2011). Transitivity of preferences. *Psychological Review, 118*(1), 42–56.

Restle, F. (1961). *Psychology of judgment and choice: A theoretical essay.* Wiley.

Rieskamp, J. (2008). The probabilistic nature of preferential choice. *Journal of Experimental Psychology: Learning, Memory, and Cognitiion, 34*(6), 1446–1465.

Rieskamp, J., & Otto, P. E. (2006). SSL: A theory of how people learn to select strategies. *Journal of Experimental Psychology: General,* 135(2), 207–236.

Rigoli, F. (2019). Reference effects on decision-making elicited by previous rewards. *Cognition, 192,* 104034.

Roe, R. M., Busemeyer, J. R., & Townsend, J. T. (2001). Multialternative decision field theory: A dynamic connectionst model of decision making. *Psychological Review, 108*(2), 370–392.

Roitman, J. D., & Shadlen, M. N. (2002). Response of neurons in the lateral intraparietal area during a combined visual discrimination reaction time task. *Journal of Neuroscience, 22*(21), 9475–9489.

Rotello, C. M., Macmillan, N. A., & Reeder, J. A. (2004). Sum-difference theory of remembering and knowing: A two-dimensional signal-detection model. *Psychological Review, 111*(3), 588–616.

Rumelhart, D. L., & Greeno, J. G. (1971). Similarity between stimuli: An experimental test of the Luce and Restle choice models. *Journal of Mathematical Psychology, 8*(3), 370–381.

Rumelhart, D. E., Hinton, G. E., & McClelland, J. L. (1986). A general framework for parallel distributed processing. *Parallel Distributed Processing: Explorations in the Microstructure of Cognition, 1,* 45–76.

Sarin, R., & Wakker, P. P. (1998). Dynamic choice and nonexpected utility. *Journal of Risk and Uncertainty, 17*(2), 87–120.

Savage, L. J. (1972/1954). *The foundations of statistics.* Courier Corporation.

Schall, J. D. (2001). Neural basis of deciding, choosing and acting. *Nature Reviews Neuroscience, 2*(1), 33–42.

Scheibehenne, B., Rieskamp, J., & González-Vallejo, C. (2009). Cognitive models of choice: Comparing decision field theory to the proportional difference model. *Cognitive Science, 33*(5), 911–939.

Schultz, W. (1998). Predictive reward signal of dopamine neurons. *Journal of Neurophysiology, 80*(1), 1–27.

Schultz, W., Dayan, P., & Montague, P. R. (1997). A neural substrate of prediction and reward. *Science, 275*(5306), 1593–1599.

Shadlen, M. N., & Kiani, R. (2013). Decision making as a window on cognition. *Neuron, 80*(3), 791–806.

Shadlen, M. N., & Newsome, W. T. (1996). Motion perception: Seeing and deciding. *Proceedings of the National Academy of Sciences, 93*(2), 628–633.

Shafir, E., & Tversky, A. (1992). Thinking through uncertainty: Nonconsequential reasoning and choice. *Cognitive Psychology, 24*(4), 449–474.

Shiffrin, R. M., & Schneider, W. (1977). Controlled and automatic human information processing: II. Perceptual learning, automatic attending and a general theory. *Psychological Review, 84*(2), 127–190.

Simon, D. A., & Daw, N. D. (2011). Neural correlates of forward planning in a spatial decision task in humans. *Journal of Neuroscience, 31*(14), 5526–5539.

Simonson, I. (1989). Choice based on reasons: The case of attraction and compromise effects. *Journal of Consumer Research, 16*(2), 158–174.

Smith, P. L. (2016). Diffusion theory of decision making in continuous report. *Psychological Review, 123*(4), 425.

Smith, S. M., & Krajbich, I. (2018). Attention and choice across domains. *Journal of Experimental Psychology: General, 147*(2), 1810–1826.

Soltani, A., De Martino, B., & Camerer, C. (2012). A range-normalization model of context-dependent choice: A new model and evidence. *PLoS Computational Biology, 8*(7), e1002607.

Solway, A., & Botvinick, M. M. (2015). Evidence integration in model-based tree search. *Proceedings of the National Academy of Sciences, 112*(37), 11708–11713.

Soto, F. A., Vucovich, L., Musgrave, R., & Ashby, F. G. (2015). General recognition theory with individual differences: A new method for examining perceptual and decisional interactions with an application to face perception. *Psychonomic Bulletin & Review, 22*(1), 88–111.

Steingroever, H., Wetzels, R., & Wagenmakers, E. J. (2016). Bayes factors for reinforcement-learning models of the Iowa gambling task. *Decision, 3*(2), 115–131.

Stewart, N. (2009). The cost of anchoring on credit-card minimum repayments. *Psychological Science, 20*(1), 39–41.

Stewart, N., & Simpson, K. (2008). A decision-by-sampling account of decision under risk. In N. Chater & M. Oaksford (Eds.), *The probabilistic mind: Prospects for Bayesian cognitive science* (pp. 261–276). Oxford University Press.

Stewart, N., Chater, N., & Brown, G. D. (2006). Decision by sampling. *Cognitive Psychology, 53*(1), 1–26.

Stewart, N., Chater, N., Stott, H. P., & Reimers, S. (2003). Prospect relativity: How choice options influence decision under risk. *Journal of Experimental Psychology: General, 132*(1), 23–46.

Stewart, N., Reimers, S., & Harris, A. J. (2014). On the origin of utility, weighting, and discounting functions: How they get their shapes and how to change their shapes. Management Science, 61(3), 687–705.

Stone, M. (1960). Models for choice-reaction time. *Psychometrika, 25*(3), 251–260.

Stott, H. P. (2006). Cumulative prospect theory's functional menagerie. *Journal of Risk and Uncertainty, 32*(2), 101–130.

Strang, G. (2016). *Introduction to linear algebra.* Wellesley-Cambridge Press.

Sun, R., & Helie, S. (2013). Psychologically realistic cognitive agents: Taking human cognition seriously. *Journal of Experimental & Theoretical Artificial Intelligence, 25*(1), 65–92.

Sutton, R. S., & Barto, A. G. (1998). *Reinforcement learning: An introduction (Vol. 1).* MIT Press.

Swets, J. A., Tanner Jr., W. P., & Birdsall, T. G. (1961). Decision processes in perception. *Psychological Review, 68*(5), 301–340.

Tanner Jr., W. P., & Swets, J. A. (1954). A decision-making theory of visual detection. *Psychological Review, 61*(6), 401–409.

Thurstone, L. L. (1927). A law of comparative judgment. *Psychological Review, 34*(4), 273–286.

Thurstone, L. L. (1959). *The measurement of values.* University of Chicago Press.

Tom, S. M., Fox, C. R., Trepel, C., & Poldrack, R. A. (2007). The neural basis of loss aversion in decision-making under risk. *Science, 315*(5811), 515–518.

Townsend, J. T., & Ashby, F. G. (1983). *Stochastic modeling of elementary psychological processes.* Columbia University Press Archive.

Train, K. E. (2009). Discrete choice methods with simulation. Cambridge University Press.

Troutman, C. M., & Shanteau, J. (1976). Do consumers evaluate products by adding or averaging attribute information? *Journal of Consumer Research, 3*(2), 101–106.

Trueblood, J. S. (2012). Multialternative context effects obtained using an inference task. *Psychonomic Bulletin & Review, 19*(5), 962–968.

Trueblood, J. S., Brown, S. D., & Heathcote, A. (2014). The multiattribute linear ballistic accumulator model of context effects in multialternative choice. *Psychological Review, 121*(2), 179–205.

Trueblood, J. S., Brown, S. D., Heathcote, A., & Busemeyer, J. R. (2013). Not just for consumers: Context effects are fundamental to decision making. *Psychological Science, 24*(6), 901–908.

Tsetsos, K., Chater, N., & Usher, M. (2012) Salience-driven value integration explains decision biases and preference reversal. *Proceedings of the National Academy of Sciences, 109,* 9659–9664.

Tsetsos, K., Usher, M., & Chater, N. (2010). Preference reversal in multiattribute choice. *Psychological Review, 117*(4), 1275–1293.

Turner, B. M. (2019). Toward a common representational framework for adaptation. *Psychological Review, 126,* 660–692.

Turner, B. M., & Van Zandt, T. (2018). Approximating Bayesian inference through model simulation. *Trends in Cognitive Sciences, 22*(9), 826–840.

Turner, B. M., Forstmann, B. U., Love, B. C., Palmeri, T. J., & Van Maanen, L. (2017). Approaches to analysis in model-based cognitive neuroscience. *Journal of Mathematical Psychology, 76,* 65–79.

Turner, B. M., Palestro, J. J., Miletic, S., and Forstmann, B. U. (2019). Advances in techniques for imposing reciprocity in brain-behavior relations. *Neuroscience and Biobehavioral Reviews, 102,* 327–336.

Turner, B. M., Schley, D. R., Muller, C., & Tsetsos, K. (2018). Competing theories of multialternative, multiattribute preferential choice. *Psychological Review, 125*(3), 329–362.

Turner, B. M., van Maanen, L., & Forstmann, B. U. (2015). Informing cognitive abstractions through neuroimaging: the neural drift diffusion model. *Psychological Review, 122*(2), 312–336.

Tversky, A. (1969). Intransitivity of preferences. *Psychological Review, 76,* 31–48.

Tversky, A. (1972a). Choice by elimination. *Journal of Mathematical Psychology, 9*(4), 341–367.

Tversky, A. (1972b). Elimination by aspects: A theory of choice. *Psychological Review, 79*(4), 281–299.

Tversky, A., & Kahneman, D. (1991). Loss aversion in riskless choice: A reference-dependent model. *Quarterly Journal of Economics, 106*(4), 1039–1061.

Tversky, A., & Kahneman, D. (1992). Advances in prospect theory: Cumulative representation of uncertainty. *Journal of Risk and Uncertainty, 5*(4), 297–323.

Tversky, A., & Russo, J. E. (1969). Substitutability and similarity in binary choices. *Journal of Mathematical Psychology, 6*(1), 1–12.

Tversky, A., & Shafir, E. (1992). The disjunction effect in choice under uncertainty. *Psychological Science, 3*(5), 305–310.

Tversky, A., & Simonson, I. (1993). Context-dependent preferences. *Management Science, 39*(10), 1179–1189.

Tversky, A., Sattath, S., & Slovic, P. (1988). Contingent weighting in judgment and choice. *Psychological Review, 95*(3), 371–384.

Tversky, A., Slovic, P., & Kahneman, D. (1990). The causes of preference reversal. *American Economic Review, 80*(1), 204–217.

Ungemach, C., Chater, N., & Stewart, N. (2009). Are probabilities overweighted or underweighted when rare outcomes are experienced (rarely)? *Psychological Science, 20*(4), 473–479.

Usher, M., & McClelland, J. L. (2001). The time course of perceptual choice: The leaky, competing accumulator model. *Psychological Review, 108*(3), 550–592.

Usher, M., & McClelland, J. L. (2004). Loss aversion and inhibition in dynamical models of multialternative choice. *Psychological Review, 111*, 757–769.

Usher, M., & Zakay, D. (1993). A neural network model for attribute-based decision processes. *Cognitive Science, 17*(3), 349–396.

Usher, M., Elhalal, A., & McClelland, J. L. (2008). The neurodynamics of choice, value-based decisions, and preference reversal. In N. Chater & Mike Oaksford (Eds.), *The probabilistic mind: Prospects for Bayesian cognitive science* (pp. 277–300). Oxford University Press.

Vandekerckhove, J., & Tuerlinckx, F. (2008). Diffusion model analysis with MATLAB: A DMAT primer. *Behavior Research Methods, 40*(1), 61–72.

Vandekerckhove, J., Tuerlinckx, F., & Lee, M. D. (2011). Hierarchical diffusion models for two-choice response times. *Psychological Methods, 16*(1), 44–62.

Van Zandt, T., & Maldonado-Molina, M. M. (2004). Response reversals in recognition memory. *Journal of Experimental Psychology: Learning, Memory, and Cognitiion, 30*(6), 1147–1166.

Verdonck, S., & Tuerlinckx, F. (2016). Factoring out nondecision time in choice reaction time data: Theory and implications. *Psychological Review, 123*(2), 208–218.

Vickers, D. (1979). *Decision processes in visual perception*. Academic Press.

Vickers, D., Smith, P., Burt, J., & Brown, M. (1985). Experimental paradigms emphasising state or process limitations: Ii effects on confidence. *Acta Psychologica, 59*(2), 163–193.

Von Neumann, J. (1955). *Mathematical foundations of quantum mechanics (No. 2)*. Princeton University Press.

Von Neumann, J., & Morgenstern, O. (2007/1945). *Theory of games and economic behavior.* Princeton University Press.

von Winterfeldt, D. (1980). Additivity and expected utility in risky multiattribute preferences. *Journal of Mathematical Psychology, 21*(1), 66–82.

Wakker, P. P. (2010). *Prospect theory: For risk and ambiguity.* Cambridge University Press.

Wald, A., & Wolfowitz, J. (1948). Optimum character of the sequential probability ratio test. *Annals of Mathematical Statistics,* 326–339.

Walsh, M. M., & Anderson, J. R. (2014). Navigating complex decision spaces: Problems and paradigms in sequential choice. *Psychological Bulletin, 140*(2), 466–486.

Wang, Z. (2014). Bridging media processing and selective exposure: A dynamic motivational model of media channel choices and choice response time. *Communication Research, 41*(8), 1064–1087.

Wang, Z., & Busemeyer, J. R. (2016). Interference effects of categorization on decision making. *Cognition, 150,* 133–149.

Wang, Z., Busemeyer, J. R., & Lang, A. (2006). Grazing or staying tuned: A dynamic stochastic model of channel changing behavior. In R. Sun & N. Miyake (Eds.), *The 28th Annual Conference of the Cognitive Science Society & the 5th International Conference of the Cognitive Science* (pp.870–875). Erlbaum.

Wedell, D. H. (1991). Distinguishing among models of contextually induced preference reversals. *Journal of Experimental Psychology: Learning, Memory, & Cognition, 17*(4), 767–778.

Wedell, D. H. (2015). Multi-alternative choice models. In G. Keren & G. Wu (Eds.), *The Wiley Blackwell handbook of judgment and decision making* (Vol. 2, 117–140.) John Wiley & Sons.

Wickens, T. D. (2002). *Elementary signal detection theory.* Oxford University Press.

Wiecki, T. V., Sofer, I., & Frank, M. J. (2013). HDDM: Hierarchical Bayesian estimation of the drift-diffusion model in Python. *Frontiers in Neuroinformatics,* 7.

Wilcox, N. T. (2011). "Stochastically more risk averse":A contextual theory of stochastic discrete choice under risk. *Journal of Econometrics, 162*(1), 89–104.

Wixted, J. T. (2007). Dual-process theory and signal-detection theory of recognition memory. *Psychological Review, 114*(1), 152–176.

Worthy, D. A., Pang, B., & Byrne, K. A. (2013). Decomposing the roles of perseveration and expected value representation in models of the Iowa Gambling Task. *Frontiers in Psychology, 4,* 640.

Yechiam, E., & Busemeyer, J. R. (2008). Evaluating generalizability and parameter consistency in learning models. *Games and Economic Behavior, 63*(1), 370–394.

Yechiam, E., Busemeyer, J. R., Stout, J. C., & Bechara, A. (2005). Using cognitive models to map relations between neuropsychological disorders and human decision-making deficits. *Psychological Science, 16*(12), 973–978.

Yellott, J. I. (1977). The relationship between Luce's choice axiom, Thurstone's theory of comparative judgment, and the double exponential distribution. *Journal of Mathematical Psychology, 15*(2), 109–144.

Yonelinas, A. P., & Parks, C. M. (2007). Receiver operating characteristics (ROCs) in recognition memory: A review. *Psychological Bulletin, 133*(5), 800–832.

Yukalov, V. I., & Sornette, D. (2011). Decision theory with prospect interference and entanglement. *Theory and Decision, 70*(3), 283–328.

Index